Operator Functions
and
Operator Equations

Operator Functions
and
Operator Equations

Michael I Gil'
Ben Gurion University of the Negev, Israel

World Scientific

W JERSEY · LONDON · SINGAPORE · BEIJING · SHANGHAI · HONG KONG · TAIPEI · CHENNAI · TOKYO

Published by

World Scientific Publishing Co. Pte. Ltd.

5 Toh Tuck Link, Singapore 596224

USA office: 27 Warren Street, Suite 401-402, Hackensack, NJ 07601

UK office: 57 Shelton Street, Covent Garden, London WC2H 9HE

British Library Cataloguing-in-Publication Data

A catalogue record for this book is available from the British Library.

OPERATOR FUNCTIONS AND OPERATOR EQUATIONS

ISBN 978-981-3221-26-0

Printed in Singapore

Preface

1. This book is devoted to norm estimates for operator-valued functions of one and two operator arguments, as well as to their applications to spectrum perturbations of operators and to linear operator equations, i.e. to equations whose solutions are linear operators. The much studied Sylvester equation is an example of such equations. Linear operator equations arise in both mathematical theory and engineering practice.

One of the first estimates for the norm of a function of a non-normal matrix was established by I.M. Gel'fand and G.E. Shilov [21] in connection with their investigations of partial differential equations. But that estimate is not sharp. It is not attained for any matrix. The problem of obtaining a sharp estimate for the norm of a matrix-valued function was repeatedly discussed in the literature, cf. [14]. In the late 1970s I obtained sharp norm estimates for the resolvents and functions of a non-normal matrix, provided the functions are regular on the convex hull of the spectrum. These estimates are attained for normal matrices. Later they were extended to various classes of non-selfadjoint operators, such as Hilbert-Schmidt's operators, operators with the Hilbert-Schmidt Hermitian components, the operators "close" to unitary ones, etc. Recently the author has derived norm estimates for some classes of functions of two operator arguments.

In the case of non-compact operators our main tool is the theory of triangular representations of operators, developed by L. de Branges, M.S. Brodskii, I.C. Gohberg, M.G. Krein, M.S. Livsic and other mathematicians.

The above mentioned results are presented in the book. We also investigate the spectral variations and rotation of invariant subspaces of operators. Besides, the well-known results of Davis and Kahan [15,16] are generalized. Moreover, we suggest bounds for the similarity condition numbers of some diagonalizable operators.

Note that our results have applications in the theories of ordinary differential [29], difference [30], functional-differential [35, 36] and integro-differential equations [41], as well as in the theories of partial integral operators [31] and analytic functions [34].

2. The aim of the book is to provide new tools for specialists in the matrix theory and functional analysis.

This is the first book that:

i) presents norm estimates for functions of two operator arguments and their applications;

ii) gives a systematic exposition of solution estimates for linear operator equations;

iii) suggests bounds for the similarity condition numbers of diagonalizable operators.

It should be noted that the results concerning the functions of one operator argument have been published in the book [28], but in the present book these results are considerably simplified and supplemented.

The book is intended not only for specialists in the matrix theory and functional analysis, but for anyone interested in various applications who has had at least a first year graduate level course in analysis.

3. The book consists of 14 chapters.

In Chapter 1, we present some well-known results from the operator theory for use in the next chapters. Here we accumulate some well-known results on operators and operator-valued functions. The material of this chapter is systematically used in the remaining chapters of the book.

Throughout the book $\mathcal{X}$ and $\mathcal{Y}$ are Banach spaces with the unit operators $I_{\mathcal{X}} = I$ and $I_{\mathcal{Y}} = I$, respectively, $\mathcal{B}(\mathcal{Y}, \mathcal{X})$ means the set of all bounded operators acting from $\mathcal{Y}$ into $\mathcal{X}$, and $\mathcal{B}(\mathcal{X}) = \mathcal{B}(\mathcal{X}, \mathcal{X})$. Chapter 2 is devoted to representations of solutions to the generalized polynomial equation

$$\sum_{j,k=0}^{m} c_{jk} A^j X B^k = C \quad (c_{jk} \in \mathbf{C}; j, k = 0, ..., m),$$

where $A \in \mathcal{B}(\mathcal{X}), B \in \mathcal{B}(\mathcal{Y})$ and $C \in \mathcal{B}(\mathcal{Y}, \mathcal{X})$ are given, and $X \in \mathcal{B}(\mathcal{Y}, \mathcal{X})$ should be found. We also investigate perturbations of the two-sided Sylvester equation

$$\sum_{j=1}^{m} A_{1j} X A_{2j} = C,$$

where $A_{1j} \in \mathcal{B}(\mathcal{X}), A_{2j} \in \mathcal{B}(\mathcal{Y})$ $(j = 1, ..., m)$.

Chapter 3 deals with functions of finite dimensional operators. In particular, we establish norm estimates for the resolvents and functions regular on the convex hull $co(A)$ of the spectrum of an operator A. We also explore the functions of the form $\frac{1}{f}(A)$, where f is regular on $co(A)$. In addition, in Chapter 3, spectrum perturbations of matrices are investigated.

In Chapter 4 we consider functions of two matrix arguments and generalized polynomial equations in Euclidean spaces.

Chapter 5 is concerned with two-sided Sylvester equations in Euclidean spaces. Solution estimates for these equations and perturbation results are obtained. These estimates enable us to obtain a bound for the distance between invariant subspaces of two matrices.

Recall that a matrix A is said to be diagonalizable, if there is an invertible matrix T, such that $T^{-1}AT = D$, where D is a normal matrix. In Chapter 6 we suggest a bound for the condition number $\kappa_T := \|T\|\|T^{-1}\|$ of a diagonalizable matrix and discuss applications of that bound to matrix functions and to equations whose coefficients are diagonalizable matrices.

Chapter 7 is devoted to functions of compact operators in a separable Hilbert space $\mathcal{H}$. Let SN_p for some $p \in [1,\infty)$ denote the Schatten-von Neumann ideal of compact operators K in $\mathcal{H}$ with the finite norm

$$N_p(K) := \sqrt[p]{\operatorname{trace}(KK^*)^{p/2}},$$

where the asterisk means the adjointness. In particular, we establish norm estimates for the resolvent of a Schatten-von Neumann operator and for functions of a Hilbert-Schmidt operator. Spectrum perturbations of compact operators and operator equations whose coefficients are compact operators are also discussed. Our inequality for the resolvent of a Schatten-von Neumann operator is deeply connected with the Carleman inequality, cf. [17].

Chapter 8 deals with non-compact non-normal operators in a separable Hilbert space. We present the triangular representations of the considered operators via the chains of their invariant projections. These representations are our main tool in the next two chapters.

In Chapter 9 we derive norm estimates for the resolvents of bounded non-selfadjoint operators A. It is supposed that either A has a Schatten-von Neumann Hermitian component, or $A^*A - I \in SN_1$. We also suggest bounds for the non-unitary eigenvalues of A in the case $A^*A - I \in SN_p$ $(p \geq 1)$.

In Chapter 10 we consider functions of a bounded non-selfadjoint operator A, assuming that the functions are regular on the convex hull of its spectrum. Besides, it is supposed that either $A - A^*$ is a Hilbert-Schmidt operator, or $AA^* - I$ is a nuclear one. Applications of the results obtained in Section 10 to the Sylvester equation are also discussed. In addition, we investigate the rotation of invariant subspaces of bounded operators with compact Hermitian components.

Chapter 11 deals with some functions of an unbounded operator in $\mathcal{H}$. In particular, we establish estimates for the resolvent and function e^{At} $(t \geq 0)$ of an unbounded operator A with a compact Hermitian component. In that chapter we also consider the Hirsch operator functions. The fraction power and operator logarithm are the examples of the Hirsch functions.

Chapter 12 is concerned with unbounded diagonalizable operators. Bounds for the condition numbers are suggested.

Let $A, B, \tilde{A} \in \mathcal{B}(\mathcal{H})$. In Chapter 13, estimates for the commutator $f(A)B - Bf(\tilde{A})$ are suggested.

Chapter 14 is devoted to a class of functions of two non-commuting operator arguments in Hilbert spaces and to the generalized polynomial equations whose coefficients are non-selfadjoint operators.

I was very fortunate to have had fruitful discussions with the late Professors I.S. Iohvidov, M.A. Krasnosel'skii and A. Pokrovskij, to whom I am very grateful for their interest in my investigations.

Michael I. Gil'

Contents

Chapter 1

Preliminaries

This chapter is of a preliminary character. Here we accumulate some well-known results on operators and operator-valued functions. The material of this chapter is systematically used in the remaining chapters of the book.

1.1 Banach and Hilbert spaces

In this section we recall very briefly some basic notions of the theory of Banach and Hilbert spaces. More details can be found in any textbook on Banach and Hilbert spaces (e.g. [1] and [18]).

Denote the set of complex numbers by $\mathbf{C}$ and the set of real numbers by $\mathbf{R}$.

A linear space $\mathcal{X}$ over $\mathbf{C}$ is called *a (complex) linear normed space* if for any $x \in \mathcal{X}$ a non-negative number $\|x\|_{\mathcal{X}} = \|x\|$ is defined, called the norm of x, having the following properties:

1. $\|x\| = 0$ iff $x = 0$,
2. $\|\alpha x\| = |\alpha| \|x\|$,
3. $\|x + y\| \le \|x\| + \|y\|$ for every $x, y \in \mathcal{X}$, $\alpha \in \mathbf{C}$.

A sequence $\{h_n\}_{n=1}^{\infty}$ of elements of $\mathcal{X}$ converges *strongly* (in the norm) to $h \in \mathcal{X}$ if

$$\lim_{n \to \infty} \|h_n - h\| = 0.$$

A sequence $\{h_n\}$ of elements of $\mathcal{X}$ is called the fundamental (Cauchy) one if

$$\|h_n - h_m\| \to 0 \text{ as } m, n \to \infty.$$

If any fundamental sequence converges to an element of $\mathcal{X}$, then $\mathcal{X}$ is called *a (complex) Banach space*.

1

Let in a linear space $\mathcal{H}$ over $\mathbf{C}$ for all $x, y \in \mathcal{H}$ a number (x, y) be defined, such that

1. $(x, x) > 0$, if $x \neq 0$, and $(x, x) = 0$, if $x = 0$,
2. $(x, y) = \overline{(y, x)}$,
3. $(x_1 + x_2, y) = (x_1, y) + (x_2, y)$ $(x_1, x_2 \in \mathcal{H})$,
4. $(\lambda x, y) = \lambda(x, y)$ $(\lambda \in \mathbf{C})$.

Then $(., .)$ is called the scalar product. Define in $\mathcal{H}$ the norm by

$$\|x\| = \sqrt{(x, x)}.$$

If $\mathcal{H}$ is a Banach space with respect to this norm, then it is called *a Hilbert space*.

The Schwarz inequality

$$|(x, y)| \leq \|x\| \, \|y\|$$

is valid.

If, in an infinite dimensional Hilbert space, there is a countable set whose closure coincides with the space, then that space is said to be *separable*. Any separable Hilbert space H possesses an orthonormal basis. This means that there is a sequence $\{e_k \in \mathcal{H}\}_{k=1}^{\infty}$ such that

$$(e_k, e_j) = 0 \text{ if } j \neq k \text{ and } (e_k, e_k) = 1 \ (j, k = 1, 2, ...),$$

and any $h \in \mathcal{H}$ can be represented as

$$h = \sum_{k=1}^{\infty} c_k e_k$$

with

$$c_k = (h, e_k), \ k = 1, 2, \ldots.$$

Besides the series strongly converges.

1.2 Linear operators

Let $Dom\,(A)$ be a subset of a Banach space $\mathcal{X}$ and A be an operator acting from $Dom\,(A)$ into a Banach space $\mathcal{Y}$. Then $Dom\,(A)$ is called the domain of A. If $Dom\,(A)$ is a linear manifold and

$$A(\alpha x_1 + \beta x_2) = \alpha A x_1 + \beta A x_2$$

for all $x_1, x_2 \in Dom\,(A)$ and $\alpha, \beta \in \mathbf{C}$, then A is called a linear operator.

If $Dom\,(A) = \mathcal{X}$ and there is a constant a, such that the inequality

$$\|Ah\|_{\mathcal{Y}} \leq a\|h\|_{\mathcal{X}} \text{ for all } h \in \mathcal{X}$$

holds, then the operator is said to be bounded. The quantity

$$\|A\|_{\mathcal{X}\to\mathcal{Y}} := \sup_{h\in\mathcal{X}} \frac{\|Ah\|_{\mathcal{Y}}}{\|h\|_{\mathcal{X}}}$$

is called the norm of A. If $\mathcal{X} = \mathcal{Y}$ we will write $\|A\|_{\mathcal{X}\to\mathcal{X}} = \|A\|_{\mathcal{X}}$ or simply $\|A\|$.

Under the natural definitions of addition and multiplication by a scalar, and the norm, *the set $\mathcal{B}(\mathcal{X},\mathcal{Y})$ of all bounded linear operators acting from $\mathcal{X}$ into $\mathcal{Y}$* becomes a Banach space. If $\mathcal{Y} = \mathcal{X}$ we will write $\mathcal{B}(\mathcal{X},\mathcal{X}) = \mathcal{B}(\mathcal{X})$. A sequence $\{A_n\} = \{A_n \in \mathcal{B}(\mathcal{X},\mathcal{Y})\}_{n=1}^{\infty}$ converges *in the uniform operator topology* (in the operator norm) to an operator $A \in \mathcal{B}(\mathcal{X},\mathcal{Y})$ if

$$\lim_{n\to\infty} \|A_n - A\|_{\mathcal{X}\to\mathcal{Y}} = 0.$$

A sequence $\{A_n\}$ *converges strongly* to an operator $A \in \mathcal{B}(\mathcal{X},\mathcal{Y})$, if the sequence of elements $\{A_n h \in \mathcal{X}\}$ strongly converges to Ah for every $h \in \mathcal{X}$.

If ϕ is a linear operator, acting from $\mathcal{X}$ into $\mathbf{C}$, then it is called a linear functional. It is bounded (continuous) if $\phi(x)$ is defined for any $x \in \mathcal{X}$, and there is a constant a such that the inequality

$$|\phi(h)| \le a\|h\|_{\mathcal{X}} \text{ for all } h \in \mathcal{X}$$

holds. The quantity

$$\|\phi\| := \sup_{h\in\mathcal{X}} \frac{|\phi(h)|}{\|h\|_{\mathcal{X}}}$$

is called *the norm of the functional* ϕ. All linear bounded functionals on $\mathcal{X}$ form a Banach space with that norm. This space is called the space *dual* to $\mathcal{X}$ and is denoted by $\mathcal{X}^*$.

In the sequel $I_{\mathcal{X}} = I$ is the identity operator in $\mathcal{X} : Ih = h$ for any $h \in \mathcal{X}$.

The operator $A^{-1} : \mathcal{Y} \to \mathcal{X}$ is the inverse one to $A \in \mathcal{B}(\mathcal{X},\mathcal{Y})$ if $AA^{-1} = I_{\mathcal{Y}}$ and $A^{-1}A = I_{\mathcal{X}}$.

Let $A \in \mathcal{B}(\mathcal{X},\mathcal{Y})$. Consider a linear bounded functional f defined on $\mathcal{Y}$. Then on $\mathcal{X}$ the linear bounded functional $g(x) = f(Ax)$ is defined. The operator A^* realizing the relation $f \to g$ is called the operator *dual (adjoint)* to A. By the definition

$$(A^*f)(x) = f(Ax) \ (x \in \mathcal{X}).$$

A^* is a bounded linear operator acting from $\mathcal{Y}^*$ to $\mathcal{X}^*$. Moreover, the relation $\|A\| = \|A^*\|$ is true.

We need the following classical result.

Theorem 1.1. *Let $\mathcal{X}, \mathcal{Y}$ be Banach spaces and let $\{A_k\}$ be a sequence of bounded linear operators on $\mathcal{X}$ to $\mathcal{Y}$. Then the limit $Ax = \lim_{n\to\infty} A_n x$ exists for every $x \in \mathcal{X}$ if and only if*
(i) the limit Ax exists for every x in a set dense in $\mathcal{X}$, and
(ii) for each $x \in \mathcal{X}$ the supremum $\sup_n \|A_n x\|_{\mathcal{X}\to\mathcal{Y}} < \infty$.
When the limit Ax exists for each $x \in \mathcal{X}$, the operator A is bounded and

$$\|A\|_{\mathcal{X}\to\mathcal{Y}} \leq \liminf_{n\to\infty} \|A_n\| \leq \sup_n \|A_n\|_{\mathcal{X}\to\mathcal{Y}}.$$

For the proof see, for instance, [18, Theorem II.3.6, p. 60].

The resolvent set $\varrho(A)$ of $A \in \mathcal{B}(\mathcal{X})$ is the set of complex numbers z for which $(A - zI)^{-1}$ exists as a bounded operator with domain $\mathcal{X}$. The spectrum $\sigma(A)$ of A is the complement of $\varrho(A)$ to the closed complex plane. The function $R_z(A) = (A - zI)^{-1}$, defined on $\varrho(A)$, is called the resolvent of A.

As it is well-known, the resolvent set is open and the resolvent is analytic in the resolvent set. In addition, $\sigma(A)$ is a closed set.

The following identity is known as the resolvent equation:

$$R_\lambda(A) - R_\mu(A) = (\lambda - \mu)R_\lambda(A)R_\mu(A) \quad (\lambda, \mu \notin \sigma(A)).$$

This equation follows by multiplying both sides of the equation

$$(\lambda I - A)(\mu I - A)[R_\lambda(A) - R_\mu(A)] = -(\mu I - A) + (\lambda I - A)$$

by $R_\lambda(A)R_\mu(A)$.

Lemma 1.1. *The spectrum of a bounded operator is bounded and non-void. Moreover*

$$\sup |\sigma(A)| = \lim_{n\to\infty} \sqrt[n]{\|A^n\|} \leq \|A\|.$$

For $|z| > \sup |\sigma(A)|$ the series

$$R_z(A) = -\sum_{k=0}^{\infty} \frac{1}{z^{k+1}} A^k$$

converges in the uniform operator topology.

For the proof see [18, Lemma V.3.4, p. 567].

The quantity

$$r_s(A) := \sup |\sigma(A)| = \lim_{n\to\infty} \sqrt[n]{\|A^n\|}$$

is called the spectral radius of A.

If there is a nontrivial solution $e \in \mathcal{X}$ of the equation $Ae = \lambda(A)e$, where $\lambda(A)$ is a number, then this number is called an eigenvalue of operator A, and e is an eigenvector corresponding to $\lambda(A)$. Any eigenvalue is a point of the spectrum.

A nonzero vector $x \in \mathcal{X}$ is called a root vector of the operator A corresponding to the eigenvalue $\lambda(A)$, if $(A - \lambda(A)I)^n x = 0$ for some positive integer n. The set of all root vectors of the operator A, corresponding to one and the same eigenvalue $\lambda(A)$ together with the zero vector, forms a lineal, which is called the root lineal. The dimension of that lineal is called the (algebraic) multiplicity of the eigenvalue $\lambda(A)$.

In the sequel $\lambda_k(A)$ $(k = 1, 2, ...)$ often denote the eigenvalues of A repeated according to their multiplicities.

An operator V is called a quasi-nilpotent one, if its spectrum consists of zero, only.

A linear operator A is called *a closed operator*, if from $x_n \in \mathcal{X} \rightarrow x_0$ and $Ax_n \rightarrow y_0$ in the norm, it follows that $x_0 \in Dom\ (A)$ and $Ax_0 = y_0$.

A bounded linear operator P is called *a projection* if $P^2 = P$.

Now let us consider operators in a Hilbert space $\mathcal{H}$. A bounded linear operator A^* is adjoint to A, if

$$(Af, g) = (f, A^* g) \text{ for every } h, g \in \mathcal{H}.$$

A linear operator A *is a selfadjoint* one, if $Dom\ (A^*) = Dom\ (A)$ and $Ax = A^* x$ $(x \in Dom\ (A))$. A selfadjoint operator A is positive (negative) definite, if

$$(Ah, h) \geq 0 \quad ((Ah, h) \leq 0) \text{ for every } h \in Dom\ (A).$$

A selfadjoint operator A is strongly positive (strongly negative) definite, if there is a constant $c > 0$, such that

$$(Ah, h) \geq c\ (h, h) \quad ((Ah, h) \leq -c\ (h, h)) \text{ for every } h \in Dom\ (A).$$

A bounded linear operator satisfying the relation $AA^* = A^* A$ is called *a normal operator*. $A \in \mathcal{B}(\mathcal{H})$ is *a unitary operator*, if $AA^* = A^* A = I$. It is clear that unitary and selfadjoint operators are examples of normal ones.

The spectrum of a selfadjoint operator is real, the spectrum of a unitary operator lies on the unit circle.

Let P be a projection in a Hilbert space. If, in addition, $P^* = P$, then it is called *an orthogonal projection (an orthoprojection)*.

1.3 Functions of a bounded operator

Let $A \in \mathcal{B}(\mathcal{X})$. By $\mathcal{F}(A)$, we denote the family of all functions which are analytic on some neighborhood of $\sigma(A)$. (The neighborhood need not be connected.)

Definition 1.1. Let $f \in \mathcal{F}(A)$ and let U be an open set whose boundary L consists of a finite number of rectifiable Jordan curves, oriented in the positive sense customary in the theory of complex variables. Suppose that $\sigma(A) \subset U$, and that $U \cup L$ is contained in the domain of analyticity of f. Then the operator $f(A)$ is defined by the equation

$$f(A) = -\frac{1}{2\pi i} \int_L f(z) R_z(A) dz.$$

Theorem 1.2. *If f, f_1 are in $\mathcal{F}(A)$, and α, β are complex numbers, then*

(a) $\alpha f + \beta f_1 \in \mathcal{F}(A)$ *and* $\alpha f(A) + \beta f_1(A) = (\alpha f + \beta f_1)(A)$;

(b) $f \cdot f_1 \in \mathcal{F}(A)$ *and* $f(A)f_1(A) = (f \cdot f_1)(A)$;

(c) if f has the power series expansion

$$f(z) = \sum_{k=0}^{\infty} c_k z^k,$$

valid in a neighborhood of $\sigma(A)$, then

$$f(A) = \sum_{k=0}^{\infty} c_k A^k.$$

Proof. In this proof we follow the proof of Theorem VII.3.10 from [18].

Statement (a) is obvious.

It is clear that $f \cdot f_1 \in \mathcal{F}(A)$. Let U_1 and U_2 be two neighborhoods of $\sigma(A)$ whose boundaries L_1 and L_2 consist of a finite number of rectifiable Jordan curves, and suppose that $U_1 \cup L_1 \subset U_2$. Suppose also that $U_2 \cup L_2$ is contained in a common region of analyticity of f and f_1. Then

$$f(A)f_1(A) = -\frac{1}{4\pi^2} \int_{L_1} f(z) R_z(A) dz \int_{L_2} f_1(z_2) R_{z_2}(A) dz_2$$

$$= -\frac{1}{4\pi^2} \int_{L_1} \int_{L_2} f(z) f_1(z_2) R_z(A) R_{z_2}(A) dz_2 \, dz$$

$$= -\frac{1}{4\pi^2} \int_{L_1} \int_{L_2} \frac{f(z)f_1(z_2)}{(z_2 - z)}(R_z(A) - R_{z_2}(A))dz\, dz_2$$

$$= -\frac{1}{4\pi^2} \int_{L_1} f(z)R_z(A) \left(\int_{L_2} \frac{f_1(z_2)}{z_2 - z}dz_2 \right) dz$$

$$+ \frac{1}{4\pi^2} \int_{L_2} f_1(z_2)R_{z_2}(A) \left(\int_{L_1} \frac{f(z)}{z_2 - z}dz \right) dz_2$$

$$= -\frac{1}{2\pi i} \int_{L_2} f(z)f_1(z)R_z(A)dz$$

$$= f(A)f_1(A)$$

by the resolvent equation and the Cauchy integral formula. This proves (b).

To prove (c), we note that the power series

$$f(z) = \sum_{k=0}^{\infty} c_k z^k$$

converges uniformly on the circle $C = \{z \in \mathbf{C} : |z| \le r_s(A) + \epsilon\}$ for ϵ sufficiently small. Consequently,

$$f(A) = -\frac{1}{2\pi i} \int_C (\sum_{k=0}^{\infty} c_k z^k)R_z(A)dz$$

$$= \frac{1}{2\pi i} \sum_{k=0}^{\infty} c_k \int_C z^k \sum_{j=0}^{\infty} \frac{A^j}{z^{j+1}}dz$$

$$= \frac{1}{2\pi i} \sum_{k=0}^{\infty} c_k \sum_{j=0}^{\infty} \int_C \frac{A^j}{z^{j-k+1}}dz$$

$$= \sum_{j=0}^{\infty} c_j A^j$$

by Cauchy's integral formula. Q.E.D.

Theorem 1.3. *(Spectral mapping theorem). If $f \in \mathcal{F}(A)$, then $\sigma(f(A)) = f(\sigma(A))$.*

Proof. In this proof we follow the proof of Theorem VII.3.11 from [18].

Let $s \in \sigma(A)$ and define the function f_1 in the domain of definition of f by

$$f_1(z) = \frac{f(s) - f(z)}{s - z}.$$

Then

$$(s - z)f_1(z) = f(s) - f(z) \in \mathcal{F}(A).$$

and $f(s)I - f(A) = (sI - A)f_1(A)$. Hence, if $f(s)I - f(A)$ has a bounded everywhere defined inverse operator A, then $f_1(A)$ would be a bounded everywhere defined inverse for $sI - A$. Consequently, $f(s) \in \sigma(f(A))$.

Conversely, let $\mu \in \sigma(f(A))$, and suppose that $f(\mu) \notin \sigma(f(A))$.

Then the function

$$h(z) = \frac{1}{f(z) - \mu}$$

belongs to $\mathcal{F}(A)$. By Theorem 1.2, $h(A)(f(A) - \mu I) = I$, which contradicts the assumption that $\mu \in \sigma(f(A))$. Q.E.D.

Theorem 1.4. *Let $f \in \mathcal{F}(A)$, $f_1 \in \mathcal{F}(f(A))$ and $F(s) = f_1(f(s))$. Then $F \in \mathcal{F}(A)$ and $F(A) = f_1(f(A))$.*

Proof. In this proof we follow the proof of Theorem VII.3.12 from [18].

The statement $F \in \mathcal{F}(A)$ follows immediately from Theorem 1.3. Let U be a neighborhood of $\sigma(f(A))$ whose boundary B consists of a finite number of rectifiable Jordan arcs, and suppose that $U \cup B$ is contained in the domain of analyticity of f_1. Let V be a neighborhood of $\sigma(A)$ whose boundary C consists of a finite number of rectifiable Jordan arcs, and suppose that $V \cup C$ is contained in the domain of analyticity of f. Suppose, moreover, that $f(V \cup C) \subseteq U$. By Theorem 1.2, the operator

$$\hat{A}(\lambda) := \frac{1}{2\pi i} \int_C \frac{(sI - A)^{-1} ds}{\lambda - f(s)}$$

satisfies the equations

$$(\lambda I - f(A))\hat{A}(\lambda) = \hat{A}(\lambda)(\lambda I - f(A)) = I.$$

Thus $\hat{A}(\lambda) = (\lambda I - f(A))^{-1}$. Consequently,

$$f_1(f(A)) = \frac{1}{2\pi i} \int_B f_1(s)(sI - f(A))^{-1} ds$$

$$= \frac{1}{4\pi^2} \int_B \int_C \frac{f_1(s) R_\lambda(A)}{s - f(\lambda)} \, d\lambda \, ds$$

$$= -\frac{1}{2\pi i} \int_C f_1(f(\lambda)) R_\lambda(A) \, d\lambda$$

by Cauchy's integral formula. Q.E.D.

Lemma 1.2. *Let $f_n \in \mathcal{F}(A), n = 1, 2, ...,$ and suppose that all the functions f_n are analytic in a fixed neighborhood V of $\sigma(A)$. Then, if $\{f_n\}$ converges uniformly to f on V, $\{f_n(A)\}$ converges to $f(A)$ in the uniform operator topology.*

Proof. In this proof we follow the proof of Lemma VII.3.13 from [18].

Let U be a neighborhood of $\sigma(A)$ whose boundary B consists of a finite number of rectifiable Jordan arcs, and such that $U \cup B \subseteq V$. Then $f_n \to f$ uniformly on B, and consequently

$$\frac{1}{2\pi i} \int_B f_n(s) R_s(A) ds$$

converges in the uniform operator topology to

$$\frac{1}{2\pi i} \int_B f(s) R_s(A) ds.$$

Q.E.D.

The next lemma can be proved in the same way, cf. [18, Lemma VII.3.14, p. 571].

Lemma 1.3. *Let V be a neighborhood of $\sigma(A)$, and let U be an open set in the complex plane. Suppose that f is an analytic function of the two complex variables s, z, for $(s, z) \in V \times U$. Then $f(A, z)$ is a $\mathcal{B}(\mathcal{X})$-valued function which is analytic for $z \in U$.*

Let us point the following result.

Lemma 1.4. *Let $A \in \mathcal{B}(\mathcal{X})$ and $f \in \mathcal{F}(A)$. Then*

$$\|f(A)\| \geq \sup_{s \in \sigma(A)} |f(s)|.$$

Proof. It is well-known that $\|A\| \geq r_s(A)$. Hence the required result is due to the Spectral Mapping theorem. Q. E. D.

In particular we have $\|A^m\| \geq r_s^m(A)$ for any integer $m \geq 1$ and $\|e^{At}\| \geq e^{\alpha(A)t}$ ($t \geq 0$), where $\alpha(A) = \sup \Re \, \sigma(A)$.

1.4   Functions of an unbounded operator

We shall now show that many of the results of the operational calculus for a bounded operator may be extended to the case of a closed operator T with a domain $Dom(T)$ and a non-empty resolvent set.

Recall that, if in the strong topology $x_n \in Dom(T), n = 1, 2, \ldots, x_n \to x$, and $Tx_n \to y$, then T is closed if $x \in Dom(T)$ and $Tx = y$.

If T is closed and everywhere defined, it is in $\mathcal{B}(\mathcal{X})$ (see [18, Section II.2.4]), hence we shall suppose that its domain is a proper subset of $\mathcal{X}$. This important case occurs for many differential operators in various function spaces. As in the case when T is in $\mathcal{B}(\mathcal{X})$, we define the resolvent set $\varrho(T)$ of T to be the set of complex numbers z such that $(Iz - T)^{-1}$ is in $\mathcal{B}(\mathcal{X})$, and the spectrum $\sigma(T)$ of T to be the complement of $\varrho(T)$ to the closed complex plane. The spectrum is divided into three disjoint sets: the point spectrum, continuous spectrum and residual spectrum. It is well-known that the spectrum is a closed set. But, in contrast to the case where T is a bounded operator, the spectrum may be a bounded set, an unbounded set, the void set, or even the whole plane, cf. [18, Exercise VII.10.1]. We exclude the last possibility, and suppose throughout this section that $\sigma(T)$ is not void. We now show how the development of an operational calculus for T may be based on the calculus already obtained for bounded operators.

Definition 1.2. For a closed operator T by $\mathcal{F}(T)$ we denote the family of all functions f which are analytic on some neighborhood of $\sigma(T)$ and at infinity.

As in the case of a bounded operator, the neighborhood need not be connected, and can depend on $f \in \mathcal{F}(T)$. Let α be a fixed point of $\varrho(T)$, and define $A = (T - \alpha I)^{-1}$. Then A defines a one-to-one mapping of $\mathcal{X}$ onto $Dom(T)$, and

$$TAx = \alpha Ax + x, x \in \mathcal{X},$$

$$ATx = \alpha Ax + x, x \in Dom(T).$$

Our objective is to define an operational calculus for T in terms of that already obtained in Section 1.3 for the bounded operator A.

If K denotes the complex sphere with its usual topology, we let $\Phi : K \to K$ be the homeomorphism defined by

$$\mu = \Phi(\lambda) = (\lambda - \alpha)^{-1}, \Phi(\infty) = 0, \Phi(\alpha) = \infty.$$

Lemma 1.5. *If $\alpha \in \varrho(T)$, then $\Phi(\sigma(T) \cup \infty) = \sigma(A)$, and the relation*

$$\phi(\mu) = f(\Phi^{-1}(\mu))$$

determines a one-to-one correspondence between $f \in \mathcal{F}(T)$ and $\phi \in \mathcal{F}(A)$.

Proof. In this proof we follow the proof of Lemma VII.9.2 from [18].
Let $\lambda \neq \alpha \in \varrho(T)$. Then $0 \neq \mu = (\lambda - \alpha)^{-1}$, and

$$(T - \alpha I)(T - \lambda I)^{-1} = [T - \lambda I + \frac{1}{\mu}I](T - \lambda I)^{-1}$$

$$= I + \frac{1}{\mu}(T - \lambda I)^{-1}.$$

But we also have

$$(T - \alpha I)(T - \lambda I)^{-1} = A^{-1}[(T - I\alpha) - \frac{1}{\mu}I]^{-1}$$

$$= \left\{ \left[(T - \alpha I) - \frac{1}{\mu}I \right] A \right\}^{-1}$$

$$= \mu(\mu I - A)^{-1},$$

which shows that

$$(T - \lambda I)^{-1} = \mu^2(\mu I - A)^{-1} - \mu I. \tag{4.1}$$

Thus $\mu \in \varrho(A)$. Conversely, if $\mu \in \varrho(A), \mu \neq 0$, then

$$(\mu I - A)^{-1}A = (A^{-1}(\mu I - A))^{-1} = (A^{-1}\mu - I)^{-1} = \frac{1}{\mu}(T - \lambda I)^{-1},$$

showing that $\lambda \in \varrho(T)$. The point $\mu = 0$ is in $\sigma(A)$, since $A^{-1} = T - \alpha I$ is unbounded. The last statement is evident from the definition of Φ. Q.E.D.

Definition 1.3. For $f \in \mathcal{F}(T)$ we define $f(T) = \phi(A)$, where $\phi \in \mathcal{F}(A)$ is given by $\phi(\mu) = f(\Phi^{-1}(\mu))$.

Theorem 1.5. *If f is in $\mathcal{F}(T)$, then $f(T)$ is independent of the choice of $\alpha \in \varrho(T)$. Let V be an open set containing $\sigma(T)$ whose boundary Γ consists of a finite number of Jordan arcs and such that f is analytic on $V \cup \Gamma$. Let Γ have positive orientation with respect to the (possibly unbounded) set V. Then*

$$f(T) = f(\infty)I - \frac{1}{2\pi i} \int_{\Gamma} f(z) R_z(T) dz.$$

For the proof see [18, Theorem VII.9.4].

Theorem 1.5 with Theorems 1.2, 1.3 and 1.4 now yields

Theorem 1.6. *Let T be a closed linear operator in $\mathcal{X}$. If f, f_1 are in $\mathcal{F}(T)$, and α, β are complex numbers, then*

(a) $\alpha f + \beta f_1 \in \mathcal{F}(T)$ *and* $\alpha f(T) + \beta f_1(T) = (\alpha f + \beta f_1)(T)$;

(b) $f \cdot f_1 \in \mathcal{F}(T)$ *and* $f(T)f_1(T) = (f \cdot f_1)(T)$;

(c) $\sigma(f(T)) = f(\sigma(T) \cup \infty)$;

(d) *if* $f \in \mathcal{F}(T), f_1 \in \mathcal{F}(f(T))$ *and* $F(s) = f_1(f(s))$,

then $F \in \mathcal{F}(T)$ *and* $F(T) = f_1(f(T))$.

1.5   The operator logarithm

The operator logarithm arises in numerous applications, in particular, its importance can be ascribed to it being the inverse function of the operator exponential. Moreover, if we consider a vector differential equation with a ω-periodic operator, then according to the Floquet theory, its Cauchy operator $U(t)$ is equal to $V(t)e^{\Gamma t}$ where $V(t)$ is a ω-periodic operator and $\Gamma = \frac{1}{\omega} \ln U(\omega)$, cf. [14].

Let $A \in \mathcal{B}(\mathcal{X})$. Assume that $0 \notin \sigma(A)$ and following the definition from the book [14, Section V.1, formula (1.6)], put

$$\ln (A) := -\frac{1}{2\pi i} \int_C \ln (z) R_z(A) dz, \qquad (5.1)$$

where the principal branch of the scalar logarithm is used, and the Jordan contour C surrounds $\sigma(A)$ and does not surround the origin.

Let us recall some additional representations of the operator logarithm. If

$$\sigma(A) \cap (-\infty, 0] = \emptyset, \qquad (5.2)$$

then

$$\ln (A) = (A - I) \int_0^\infty (tI + A)^{-1} \frac{dt}{1 + t}. \qquad (5.3)$$

Indeed,

$$\int_0^\infty (t + x)^{-1} \frac{dt}{1 + t} (x - 1) = -\lim_{a \to \infty} \int_0^a \left[\frac{1}{t + x} - \frac{1}{1 + t} \right] dt = \ln \ x.$$

$$(x \notin (-\infty, 0]).$$

Hence we have (5.3) for finite dimensional operators with the simple spectra. Since the set of matrices with simple spectra is dense in the set of all matrices, (5.3) is valid for all matrices. In the general situation formula (5.3) has been proved in [69, Section 10.1].

Recall that $r_s(.)$ means the spectral radius. If

$$r_s(I - A) < 1, \tag{5.4}$$

then one can use the obvious representation

$$\ln(A) = \sum_{k=1}^{\infty} \frac{1}{k}(I - A)^k. \tag{5.5}$$

1.6   Perturbations of operators in uniform topology

Let $A, B \in \mathcal{B}(\mathcal{X})$. Since

$$R_\lambda(A) - R_\lambda(B) = R_\lambda(A)(B - \lambda I - (A - \lambda I))R_\lambda(B),$$

we get *the Hilbert identity for resolvents*

$$R_\lambda(A) - R_\lambda(B) = -R_\lambda(A)(A - B)R_\lambda(B) \quad (\lambda \notin \sigma(A) \cup \sigma(B)). \tag{6.1}$$

Let R_0 be a set in the complex plane and let $\epsilon > 0$. By $S(R_0, \epsilon)$ we denote the ϵ-neighborhood of R_0. That is,

$$\text{dist}\{R_0, S(R_0, \epsilon)\} \le \epsilon.$$

The Hilbert identity for resolvents enables us to prove the following well-known result.

Lemma 1.6. *Let $A \in \mathcal{B}(\mathcal{X})$ and let $\epsilon > 0$. Then there is a $\delta > 0$, such that, if a bounded operator B in $\mathcal{X}$ satisfies the condition*

$$\|A - B\| \le \delta, \tag{6.2}$$

then $\sigma(B)$ lies in $S(\sigma(A), \epsilon)$ and

$$\|R_\lambda(A) - R_\lambda(B)\| \le \epsilon \tag{6.3}$$

for any λ, which does not belong to $S(\sigma(A), \epsilon)$.

For the details of the proof of this lemma we refer the reader to the book [18, p. 585, Lemma VII.6.3].

Lemma 1.7. *Let $A \in \mathcal{B}(\mathcal{X})$ and let $\epsilon > 0$. Then there is a $\delta > 0$, such that, if a bounded operator B in $\mathcal{X}$ satisfies condition (6.2) and f is regular on a neighborhood W of $\sigma(A) \cup \sigma(B)$, then*

$$\|f(A) - f(B)\| \le \epsilon.$$

Proof. Due to (6.1),

$$f(A) - f(B) = -\frac{1}{2\pi i} \int_L f(z)(R_z(A) - R_z(B))dz =$$

$$\frac{1}{2\pi i} \int_L f(z)(R_z(A)(A - B)R_z(B)dz,$$

where $L \subset W$ is a closed Jordan contour surrounding $\sigma(A) \cup \sigma(B)$. So

$$\|f(A) - f(B)\| \leq \|A - B\|\frac{1}{2\pi} \int_L |f(z)|\|R_z(A)\|\|R_z(B)\||dz|.$$

But due to the previous lemma, for given ϵ_1, $\|R_z(B)\| \leq (\|R_z(A)\| + \epsilon_1)$, provided (6.2) holds. This proves the lemma. Q. E. D.

We need also the following well-known result proved in [17, Lemma XI.9.5].

Lemma 1.8. *Let A, A_n $(n = 1, 2, ...)$ be compact operators in $\mathcal{X}$, and let $A_n \to A$ in the uniform operator topology. Let $\lambda_m(A)$ be an enumeration of the non-zero eigenvalues of A, each repeated according to its multiplicity. Then there exist enumerations $\lambda_m(A_n)$ of the non-zero eigenvalues of A_n, with repetitions according to multiplicity, such that*

$$\lim_{n \to \infty} \lambda_m(A_n) = \lambda_m(A) \ , m \geq 1,$$

the limit being uniform in m.

If a sequence of operators $A_n \in \mathcal{B}(\mathcal{X})$ $(n = 1, 2, ...)$ converges in the operator norm to an operator $A \in \mathcal{B}(\mathcal{X})$, then due to the upper semi continuity of the spectrum [60, p. 208], [58, p. 56, Problem 103],

$$\lim_{n \to \infty} \sigma(A_n) \subseteq \sigma(A). \tag{6.4}$$

1.7 Perturbations of operators in strong topology

Throughout this book the symbol $A_n \overset{s}{\to} A$ means that a sequence of operators $A_n \in \mathcal{B}(\mathcal{X})$ $(n = 1, 2, ...)$ strongly converges to $A \in \mathcal{B}(\mathcal{X})$ as $n \to \infty$. If $A_n \overset{s}{\to} A$, then Theorem 1.1 yields the inequality

$$\|A\| \leq \lim_{n \to \infty} \inf \|A_n\|. \tag{7.1}$$

Lemma 1.9. *Let $A_n \overset{s}{\to} A$ and a $\lambda \in \mathbf{C}$ be a regular point for all A_n and A. In addition, let $\sup_n \|R_\lambda(A_n)\| < \infty$. Then $R_\lambda(A_n) \overset{s}{\to} R_\lambda(A)$, and therefore,*

$$\|R_\lambda(A)\| \leq \lim_{n \to \infty} \inf \|R_\lambda(A_n)\|. \tag{7.2}$$

Proof. Making use of identity (6.1), we have

$$\|R_\lambda(A)x - R_\lambda(A_n)x\| \leq \|R_\lambda(A_n)\|\|(A - A_n)R_\lambda(A)x\| \leq$$

$$\sup_n \|R_\lambda(A_n)\|\|(A - A_n)R_\lambda(A)x\| \to 0 \quad (x \in \mathcal{X}).$$

This proves the required result. Q. E. D.

Denote

$$\rho(A, \lambda) := \inf_{s \in \sigma(A)} |\lambda - s|.$$

Let

$$\|R_\lambda(A_n)\| \leq F\left(\frac{1}{\rho(A_n, \lambda)}\right) \quad (\lambda \notin \sigma(A_n); n = 1, 2, ...), \qquad (7.3)$$

where $F(t)$ is a monotonically increasing non-negative continuous function of a non-negative variable t, independent of n, such that $F(0) = 0$ and $F(\infty) = \infty$. Assume that

$$\sigma(A_n) \subseteq \sigma(A) \quad (n = 1, 2, ...). \qquad (7.4)$$

Then $\rho(A_n, \lambda) \geq \rho(A, \lambda)$ and

$$\|R_\lambda(A_n)\| \leq F\left(\frac{1}{\rho(A, \lambda)}\right) \quad (\lambda \notin \sigma(A); n = 1, 2, ...).$$

The right-hand part of this inequality does not depend on n. Now the previous lemma implies

Corollary 1.1. *Let a sequence of operators $A_n \in \mathcal{B}(\mathcal{X})$ $(n = 1, 2, ...)$ strongly converge to an operator $A \in \mathcal{B}(\mathcal{X})$. Let conditions (7.3) and (7.4) hold. Then*

$$\|R_\lambda(A)\| \leq F\left(\frac{1}{\rho(A, \lambda)}\right) \quad (\lambda \notin \sigma(A)).$$

Moreover, the following result is valid.

Corollary 1.2. *Let a sequence of operators $A_n \in \mathcal{B}(\mathcal{X})$ $(n = 1, 2, ...)$ strongly converge to $A \in \mathcal{B}(\mathcal{X})$ and satisfy conditions (7.3) and (7.4). Let f be regular on a simply connected open set $M \supset \sigma(A)$. Then the sequence of operators $f(A_n)$ $(n = 1, 2, ...)$ strongly converges to $f(A)$ and*

$$\|f(A)\| \leq \lim_{n \to \infty} \inf \|f(A_n)\|. \qquad (7.5)$$

Indeed, due to (6.1),

$$(f(A) - f(A_n))x =$$

$$\frac{1}{2\pi i} \int_C f(z) R_z(A_n)(A_n - A) R_z(A)x \ dz \quad (x \in \mathcal{X}), \tag{7.6}$$

where $C \subset M$ is a closed Jordan contour surrounding $\sigma(A)$. Due to the previous corollary,

$$\|R_z(A_n)(A - A_n)R_z(A)x\| \to 0 \quad (z \in C).$$

From (7.3), (7.4) and Corollary 1.1 it follows

$$\sup_{z \in C, n=1,2,\ldots} \|R_z(A_n)(A - A_n)R_z(A)x\|$$

$$\leq \|x\| F^2(1/\rho(A,z)) \sup_{n=1,2,\ldots} \|A - A_n\| < \infty.$$

So by the Lebesgue theorem

$$\int_C |f(z)| \|R_z(A_n)(A - A_n)R_z(A)x\| |dz| \to 0.$$

According to (7.6), $f(A_n) \overset{s}{\to} f(A)$ and (7.1) yields the assertion of Corollary 1.2. Q. E. D.

In the following corollary we do not require conditions (7.3) and (7.4).

Corollary 1.3. *Let a sequence of operators $A_n \in \mathcal{B}(\mathcal{X})$ $(n = 1, 2, \ldots)$ converge strongly to $A \in \mathcal{B}(\mathcal{X})$. Let f be regular on the disc $\{z \in \mathbf{C} : |z| \leq m_0\}$ with $m_0 > \nu_0 := \sup_n \|A_n\|$. Then $\{f(A_n)\}$ strongly converges to $f(A)$ and (7.5) holds.*

Indeed, for any λ with $|\lambda| > m_0$ we have $\|(A_n - \lambda)^{-1}\| \leq 1/(|\lambda| - m_0)^{-1}$. Then due to Lemma 1.9, $R_\lambda(A_n) \overset{s}{\to} R_\lambda(A)$, and thus, according to (7.6) with $C = \{z \in \mathbf{C} : |z| = m_0\}$, for any $x \in \mathcal{X}$ we have

$$\|(f(A) - f(A_n))x\| \leq \frac{1}{2\pi(m_0 - \nu_0)} \int_{|z|=m_0} |f(z)| \|(A_n - A)R_z(A)x\| dz \to 0.$$

Q. E. D.

1.8 Spectral variations

Let $A, \tilde{A} \in \mathcal{B}(\mathcal{X})$. Put $q := \|A - \tilde{A}\|$. Due to the Hilbert identity (6.1),

$$\|R_\lambda(\tilde{A})\| \le \|R_\lambda(A)\| + q\|R_\lambda(A)\|\|R_\lambda(\tilde{A})\|.$$

So if a $\lambda \in \mathbb{C}$ is regular for A and

$$q\|R_\lambda(A)\| < 1, \tag{8.1}$$

then λ is also regular for $\tilde{A}$. Moreover,

$$\|R_\lambda(\tilde{A})\| \le \frac{\|R_\lambda(A)\|}{1 - q\|R_\lambda(A)\|}. \tag{8.2}$$

Definition 1.4. The quantity

$$\mathrm{sv}_A(\tilde{A}) := \sup_{\mu \in \sigma(\tilde{A})} \inf_{\lambda \in \sigma(A)} |\mu - \lambda|$$

is called the spectral variation off $\tilde{A}$ with respect to A. In addition,

$$hd(A, \tilde{A}) := \max\{\mathrm{sv}_A(\tilde{A}), \mathrm{sv}_{\tilde{A}}(A)\}$$

is the Hausdorff distance between the spectra of A and $\tilde{A}$.

Again assume that

$$\|R_\lambda(A)\| \le F\left(\frac{1}{\rho(A, \lambda)}\right) \quad (\lambda \notin \sigma(A)), \tag{8.3}$$

where $F(t)$ is a monotonically increasing non-negative continuous function of a non-negative variable t, independent of n, such that $F(0) = 0$ and $F(\infty) = \infty$. We need the following technical lemma.

Lemma 1.10. *Let $A, \tilde{A} \in \mathcal{B}(\mathcal{X})$ and condition (8.3) hold. Then $\mathrm{sv}_A(\tilde{A}) \le z(F, q)$, where $z(F, q)$ is the unique positive root of the equation*

$$qF(1/z) = 1. \tag{8.4}$$

Proof. Due to (8.1)

$$1 \le qF\left(\frac{1}{\rho(A, \mu)}\right) \quad \text{for all } \mu \in \sigma(\tilde{A}).$$

Compare this inequality with (8.4). Since $F(t)$ monotonically increases, $z(F, q)$ is a unique positive root of (8.3) and $\rho(A, \mu) \le z(F, q)$. This proves the required result. Q. E. D.

1.9   Rotations of simple eigenvectors

Recall that an eigenvalue is said to be simple, if its multiplicity is equal to one. In this section we show that norm estimates for resolvents give us a possibility to derive the norm of the difference of eigenvectors corresponding to simple eigenvalues of two operators.

For a $b \in \mathbf{C}$ and an $r > 0$, put

$$\Omega(b, r) := \{z \in \mathbf{C} : |z - b| < r\}.$$

Let A and $\tilde{A}$ be bounded linear operators acting in a Banach space $\mathcal{X}$, and $q = \|A - \tilde{A}\|$. Suppose that

$$A \text{ has an isolated simple eigenvalue } \mu. \tag{9.1}$$

So

$$\chi(A) := \frac{1}{2} \inf_{s \in \sigma(A), s \neq \mu} |s - \mu| > 0.$$

Lemma 1.11. *Let condition (9.1) hold and for a positive* $r < \chi(A, \mu)$, *let*

$$q \sup_{|z - \mu| = r} \|R_z(A)\| < 1. \tag{9.2}$$

Then $\tilde{A}$ *has in* $\Omega(\mu, r)$ *a simple eigenvalue.*

This result is a particular case of the well-known Theorem IV.3.18 from [60].

Under condition (9.1) assume that

$$\|R_\lambda(A)\| \leq F(1/|\lambda - \mu|) \quad (|\lambda - \mu| \leq \chi(A), \lambda \neq \mu), \tag{9.3}$$

where $F(x)$ again is a monotonically increasing continuous function of $x \geq 0$, such that $F(0) = 0$ and $F(\infty) = \infty$.

Put $\partial\Omega := \{z \in \mathbf{C} : |z - \mu| = \chi(A)\}$,

$$P(A) = -\frac{1}{2\pi i} \int_{\partial\Omega} R_\lambda(A) d\lambda \text{ and } P(\tilde{A}) = -\frac{1}{2\pi i} \int_{\partial\Omega} R_\lambda(\tilde{A}) d\lambda.$$

That is, $P(A)$ and $P(\tilde{A})$ are the projections onto the eigen-spaces of A and $\tilde{A}$, respectively, corresponding to points of the spectra, which lie in $\Omega(\mu, \chi(A))$.

Lemma 1.12. *Under conditions (9.1) and (9.3), let*

$$qF(1/\chi(A)) < 1. \tag{9.4}$$

Then $\|P(A) - P(\tilde{A})\| \leq \xi(q, A)$ *where*

$$\xi(q, A) := \frac{q\chi(A)F^2(1/\chi(A))}{1 - qF(1/\chi(A))}.$$

Proof. Due to the previous lemma dim $P(A)\mathcal{X} = \dim P(\tilde{A})\mathcal{X} = 1$. From (9.3) and (9.4) it follows that

$$\|R_\lambda(\tilde{A})\| \leq \|R_\lambda(A)\|(1 - qF(1/\chi(A)))^{-1} \leq \frac{F(1/\chi(A))}{1 - qF(1/\chi(A))} \quad (\lambda \in \partial\Omega)$$

and therefore

$$\|P(A) - P(\tilde{A})\| \leq \frac{1}{2\pi}\int_{\partial\Omega}\|R_\lambda(A) - R_\lambda(\tilde{A})\||d\lambda| \leq$$

$$\frac{1}{2\pi}\int_{\partial\Omega}\|R_\lambda(\tilde{A})\|qF(1/\chi(A))|d\lambda| \leq \frac{qF^2(1/\chi(A))\chi(A)}{1 - qF(1/\chi(A))},$$

as claimed. Q. E. D.

Lemma 1.13. *Let Q_1 and Q_2 be projections in $\mathcal{X}$, satisfying the condition $r := \|Q_1 - Q_2\| < 1$. Then for any eigenvector f_1 of Q_1 with $\|f_1\| = 1$ and $Q_1 f_1 = f_1$, there exists an eigenvector f_2 of Q_2 with $\|f_2\| = 1$ and $Q_2 f_2 = f_2$, such that*

$$\|f_1 - f_2\| \leq \frac{2r}{1-r}.$$

Proof. We have $\|Q_2 f_1 - Q_1 f_1\| \leq r < 1$ and

$$b_0 := \|Q_2 f_1\| \geq \|Q_1 f_1\| - \|(Q_1 - Q_2)f_1\| \geq 1 - r > 0.$$

Thanks to the relation $Q_2(Q_2 f_1) = Q_2 f_1$, $Q_2 f_1$ is an eigenvector of Q_2. Then $f_2 = b_0^{-1}\|Q_2 f_1\|$ is a normed eigenvector of Q_2. So

$$f_1 - f_2 = Q_1 f_1 - b_0^{-1} Q_2 f_1 = f_1 - b_0^{-1} f_1 + b_0^{-1}(Q_1 - Q_2)f_1.$$

But

$$\frac{1}{b_0} \leq \frac{1}{1-r}$$

and

$$\|f_1 - f_2\| \leq (b_0^{-1} - 1)\|f_1\| + b_0^{-1}\|(Q_1 - Q_2)f_1\|$$

$$\leq \frac{1}{1-r} - 1 + \frac{r}{1-r} = \frac{2r}{1-r},$$

as claimed. Q. E. D.

Lemma 1.14. *Let the conditions (9.1), (9.3) and $\xi(q, A) < 1$ hold. Then for the normed eigenvector e of A corresponding to μ, there exists a normed eigenvector $\tilde{e}$ of $\tilde{A}$, corresponding to the simple eigenvalue $\lambda(\tilde{A}) \in \Omega(\mu, \chi(A))$ such that*

$$\|e - \tilde{e}\| \leq \frac{2\xi(q, A)}{1 - \xi(q, A)}.$$

Proof. Due to Lemma 1.12, we obtain

$$\|P(A) - P(\tilde{A})\| \leq \xi(q, A) < 1.$$

Now the required result is due to the previous lemma. Q. E. D.

1.10   Comments to Chapter 1

The chapter contains mostly well-known results. Corollaries 1.1 and 1.2 are probably new. Lemma 1.14 is taken from [28].

Chapter 2

Representations of Solutions to Operator Equations

This chapter is devoted to representations of solutions to some classes of operator equations in Banach spaces.

Throughout this chapter $\mathcal{X}$ and $\mathcal{Y}$ are Banach spaces with the unit operators $I_{\mathcal{X}} = I$ and $I_{\mathcal{Y}} = I$, respectively. Recall that $\mathcal{B}(\mathcal{Y}, \mathcal{X})$ means the set of all bounded operators acting from $\mathcal{Y}$ into $\mathcal{X}$ and $\mathcal{B}(\mathcal{X}) = \mathcal{B}(\mathcal{X}, \mathcal{X})$.

Let $A \in \mathcal{B}(\mathcal{X}), B \in \mathcal{B}(\mathcal{Y})$ and $C \in \mathcal{B}(\mathcal{Y}, \mathcal{X})$ be given. Then the equations

$$\sum_{j=0}^{m} c_j A^{m-j} X B^j = C$$

and

$$\sum_{j=0}^{m} c_j A^j X B^j = C \ \ (c_j \in \mathbf{C}; j = 0, ..., m < \infty),$$

where $X \in \mathcal{B}(\mathcal{Y}, \mathcal{X})$ should be found, will be called *the polynomial operator equations*. The equation

$$\sum_{j,k=0}^{m} c_{jk} A^j X B^k = C \ \ (c_{jk} \in \mathbf{C}, j, k = 0, ..., m)$$

will be called *the generalized polynomial equation*. The following equations are important examples of the polynomial operator equations: the Sylvester equation

$$AX - XB = C$$

and the equation

$$X - AXB = C,$$

which will be called the *quasi-Sylvester equation*.

Furthermore, the equation

$$\sum_{j=1}^{m} A_{1j} X A_{2j} = C,$$

where $A_{1j} \in \mathcal{B}(\mathcal{X}), A_{2j} \in \mathcal{B}(\mathcal{Y})$ $(j = 1, ..., m)$ are given operators, will be called *the two-sided Sylvester equation.*

2.1　Generalized polynomial operator equations

In this section we follow the book [14, Chapter 1].

Again consider the generalized polynomial equation

$$\sum_{j,k=0}^{m} c_{jk} A^j X B^k = C \quad (A \in \mathcal{B}(\mathcal{X}), B \in \mathcal{B}(\mathcal{Y}), C \in \mathcal{B}(\mathcal{Y}, \mathcal{X})). \tag{1.1}$$

Let A_l and B_r denote the linear operators acting on any operator $Z \in \mathcal{B}(\mathcal{Y}, \mathcal{X})$ that are induced by multiplying Z on the left by A and on the right by B, respectively:

$$A_l Z := AZ \quad \text{and} \quad B_r Z := ZB.$$

It is easily seen that the operators A_l and B_r commute. With the use of the polynomial

$$P(\lambda, \mu) = \sum_{j,k=0}^{m} c_{jk} \lambda^j \mu^k \quad (\lambda, \mu \in \mathbf{C})$$

we form the operator

$$P_{A,B} = P(A_l, B_r) := \sum_{j,k=0}^{m} c_{jk} A_l^j B_r^k. \tag{1.2}$$

Equation (1.1) now takes the form $P_{A,B} X = C$.

We will first determine a condition for the existence of the inverse of $P_{A,B}$. Suppose $\lambda \notin \sigma(A)$. Then the operator

$$(A_l - \lambda I)^{-1} = (A - \lambda I)_l^{-1}$$

defined by $(A - \lambda I)_l^{-1} Z = (A - \lambda I)^{-1} Z$ clearly exists. This means in particular that $\sigma(A_l) \subseteq \sigma(A)$. (It can in fact be shown that $\sigma(A_l) = \sigma(A_r) = \sigma(A)$.) In exactly the same way $\sigma(B_r) \subseteq \sigma(B)$.

We can now explicitly express the operator $P(A_l, B_r)$ in terms of the polynomial $P(\lambda, \mu)$. Indeed, use the formula for the powers:

$$A^j = -\frac{1}{2\pi i} \int_{\Gamma_A} \lambda^j R_\lambda(A) d\lambda, \, B^j = -\frac{1}{2\pi i} \int_{\Gamma_B} \lambda^j R_\lambda(B) d\lambda \quad (j = 1, 2, ...),$$

where Γ_A and Γ_B are closed Jordan contours containing $\sigma(A)$ and $\sigma(B)$, respectively. By substituting these expressions into (1.2), we have

$$P(A_l, B_r) = -\frac{1}{4\pi^2} \int_{\Gamma_A} \int_{\Gamma_B} P(\lambda, \mu)(A_l - \lambda I)^{-1}(B_r - \mu I)^{-1} d\lambda \, d\mu. \quad (1.3)$$

Formula (1.3) suggests the following generalization of the problem under consideration. Let $K_{A,B}$ be the set of all functions $\phi(\lambda, \mu)$ that are regular on a neighborhood of the set $\sigma(A) \times \sigma(B)$. For each function $\phi(\lambda, \mu) \in K_{A,B}$ we define the operator $\phi_{A,B} = \phi(A_l, B_r)$ acting from $\mathcal{Y}$ into $\mathcal{X}$ by

$$\phi_{A,B} Z = \phi(A_l, B_r) Z$$

$$= -\frac{1}{4\pi^2} \int_{\Gamma_A} \int_{\Gamma_B} \phi(\lambda, \mu)(A - \lambda I)^{-1} Z (B - \mu I)^{-1} d\lambda \, d\mu \quad (1.4)$$

for all $Z \in \mathcal{B}(\mathcal{Y}, \mathcal{X})$, or simply

$$\phi(A_l, B_r) = -\frac{1}{4\pi^2} \int_{\Gamma_A} \int_{\Gamma_B} \phi(\lambda, \mu)(A_l - \lambda I)^{-1}(B_r - \mu I)^{-1} d\lambda \, d\mu. \quad (1.5)$$

Formula (1.4) establishes a correspondence, between the functions of class $K_{A,B}$ and a certain commutative set of operators from $\mathcal{B}(\mathcal{Y}, \mathcal{X})$ that has the following properties:

a) If $\phi(\lambda, \mu) \equiv 1$ then $\phi(A_l, B_r) = I$.

b) If $\phi(\lambda, \mu) = a_1 \phi_1(\lambda, \mu) + a_2 \phi_2(\lambda, \mu)$ $(a_1, a_2 = constant)$, then

$$\phi(A_l, B_r) = a_1 \phi_1(A_l, B_r) + a_2 \phi_2(A_l, B_r).$$

c) If $\phi(\lambda, \mu) = \phi_1(\lambda, \mu)\phi_2(\lambda, \mu)$, then, as follows from Theorem 1.2,

$$\phi(A_l, B_r) = \phi_1(A_l, B_r)\phi_2(A_l, B_r).$$

d) If $\lim_{n \to \infty} \phi_n(\lambda, \mu) = \phi(\lambda, \mu)$

uniformly in a neighborhood of the Cartesian product $\sigma(A) \times \sigma(B)$, then

$$\lim_{n \to \infty} \phi_n(A_l, B_r) = \phi(A_l, B_r).$$

The above remarks imply the following proposition.

Theorem 2.1. *Suppose a function $\phi(\lambda, \mu)$ of class $K_{A,B}$ does not vanish for*

$$(\lambda, \mu) \in \sigma(A) \times \sigma(B).$$

Then the operator $\phi(A_l, B_r)$ has an inverse $\psi(A_l, B_r)$ with

$$\psi(\lambda, \mu) = \frac{1}{\phi(\lambda, \mu)}.$$

In other words, the equation

$$\phi(A_l, B_r)X = C \tag{1.6}$$

has in this case a unique solution $X \in \mathcal{B}(\mathcal{Y}, \mathcal{X})$ for each $C \in \mathcal{B}(\mathcal{Y}, \mathcal{X})$. It is representable in the form

$$X = -\frac{1}{4\pi^2} \int_{\Gamma_A} \int_{\Gamma_B} \frac{1}{\phi(\lambda, \mu)} (A - \lambda I)^{-1} C (B - \mu I)^{-1} d\lambda \, d\mu. \tag{1.7}$$

We formulate this result for equation (1.1) in particular.

Theorem 2.2. *If the condition*

$$P(\lambda, \mu) = \sum_{j,k=0}^{m} c_{jk} \lambda^j \mu^k \neq 0 \quad ((\lambda, \mu) \in \sigma(A) \times \sigma(B)) \tag{1.8}$$

is satisfied, then equation (1.1) has for each $C \in \mathcal{B}(\mathcal{Y}, \mathcal{X})$ a unique solution $X \in \mathcal{B}(\mathcal{Y}, \mathcal{X})$ which is representable in the form

$$X = -\frac{1}{4\pi^2} \int_{\Gamma_A} \int_{\Gamma_B} \frac{1}{P(\lambda, \mu)} (A - \lambda I)^{-1} C (B - \mu I)^{-1} d\lambda \, d\mu.$$

2.2   The quasi-Sylvester equation

Consider the equation

$$Y - AYB = C \quad (A \in \mathcal{B}(\mathcal{X}), B \in \mathcal{B}(\mathcal{Y}), C \in \mathcal{B}(\mathcal{Y}, \mathcal{X})). \tag{2.1}$$

According to (1.2) it can be written as

$$(I - A_l B_r)Y = C.$$

So

$$Y = (I - A_l B_r)^{-1} C, \tag{2.2}$$

provided

$$\lambda\mu \neq 1 \quad ((\lambda, \mu) \in \sigma(A) \times \sigma(B))$$

Let us assume that the spectral radiuses of A and B satisfy the condition

$$r_s(A) r_s(B) < 1. \tag{2.3}$$

Lemma 2.1. *Let condition (2.3) hold. Then equation (2.1) has a unique solution Y, which can be represented as*

$$Y = \sum_{k=0}^{\infty} A^k C B^k, \tag{2.4}$$

where the series converges in the operator norm. Moreover, for any $\epsilon > 0$ satisfying the condition $(r_s(A) + \epsilon)(r_s(B) + \epsilon) < 1$, there is a constant c_ϵ, such that

$$\|Y\| \leq \frac{\|C\| c_\epsilon}{1 - (r_s(A) + \epsilon)(r_s(B) + \epsilon)}.$$

Proof. The existence and uniqueness of solutions is due to Theorem 2.2, since $P(\lambda, \mu) = 1 - \lambda\mu \neq 0$ $((\lambda, \mu) \in \sigma(A) \times \sigma(B))$ under consideration. Due to the representation

$$A^k = -\frac{1}{2\pi i} \int_{|\lambda| = r_1} \lambda^k R_\lambda(A) d\lambda \quad (r_1 > r_s(A); \; k = 1, 2, ...)$$

we can assert that for any $\epsilon > 0$, there is a constant c_A, such that

$$\|A^k\| \leq c_A (r_s(A) + \epsilon)^k.$$

Similarly,

$$\|B^k\| \leq c_B (r_s(B) + \epsilon)^k \quad (c_B = const).$$

We have

$$\|A^k\| \|B^k\| \leq c_A \, c_B \delta^k$$

with $\delta = (r_s(A) + \epsilon)(r_s(B) + \epsilon) < 1$. So

$$\sum_{k=0}^{\infty} \|A^k\| \|B^k\| \leq c_A c_B \frac{1}{1 - \delta}.$$

From (2.4) we deduce that

$$Y - AYB = \sum_{k=0}^{\infty} A^k C B^k - \sum_{k=0}^{\infty} A^{k+1} C B^{k+1} = C,$$

proving (2.4). Moreover,

$$\|Y\| \leq \|C\| \sum_{k=0}^{\infty} \|A^k\| \|B^k\| \leq c_A c_B \|C\| \frac{1}{1 - \delta}.$$

This finishes the proof. Q. E. D.

If (2.3) holds, then obviously, there is a $b \in \mathbf{C}, b \neq 0$, such that

$$|b| r_s(A) < 1 \text{ and } r_s(B) < |b|. \tag{2.5}$$

Conversely, if (2.5) holds, then (2.3) is valid.

Lemma 2.2. *Let a complex number $b \neq 0$ satisfy condition (2.5). Then (2.1) has the unique solution Y which is representable as*

$$Y = \frac{b}{2\pi} \int_0^{2\pi} (Ie^{-i\omega} - bA)^{-1} C (be^{i\omega} I - B)^{-1} d\omega.$$

Proof. Clearly,

$$(Ie^{-i\omega} - bA)^{-1}C(Ie^{i\omega} - \frac{1}{b}B)^{-1} =$$

$$(I - e^{i\omega}bA)^{-1}C(I - e^{-i\omega}\frac{1}{b}B)^{-1} =$$

$$\sum_{k=0}^{\infty}\sum_{j=0}^{\infty}(be^{i\omega}A)^k Ce^{-ij\omega}(\frac{1}{b}B)^j.$$

But

$$\int_0^{2\pi} e^{(k-j)i\omega}d\omega = 0 \ \ (k \neq j).$$

So due to the previous lemma

$$\int_0^{2\pi} (Ie^{-i\omega} - bA)^{-1}C(Ie^{i\omega} - \frac{1}{b}B)^{-1}d\omega =$$

$$2\pi \sum_{k=0}^{\infty} A^k CB^k = 2\pi Y,$$

as claimed. Q. E. D.

An additional representation for solutions of (2.1) is given in Section 2.4 below.

2.3   The Sylvester equation

Let us consider the equation

$$AX - XB = C. \tag{3.1}$$

In this case $P(\lambda, \mu) = \lambda - \mu$. According to (1.2), equation (3.1) can be written as

$$(A_l - B_r)X = C.$$

Thus due to Theorem 2.2, under the condition

$$\sigma(A) \cap \sigma(B) = \emptyset, \tag{3.2}$$

equation (3.1) has a unique solution, and it is given by the formula

$$X = (A_l - B_r)^{-1}C$$

$$= -\frac{1}{4\pi^2} \int_{\Gamma_A} \int_{\Gamma_B} \frac{1}{\lambda - \mu}(A - \lambda I)^{-1}C(B - \mu I)^{-1}d\lambda \, d\mu. \qquad (3.3)$$

The next theorem also gives an expression for the solution of (3.1) whenever $\sigma(A)$ and $\sigma(B)$ are disjoint, without any more special assumptions about the separation of the spectra.

Theorem 2.3. *(Rosenblum [75]). Let condition (3.2) hold and Γ be a union of closed contours in the plane, with total winding numbers 1 around $\sigma(A)$ and 0 around $\sigma(B)$. Then the solution of equation (3.1) can be expressed as*

$$X = \frac{1}{2\pi i} \int_\Gamma (A - \lambda I)^{-1}C(B - \lambda I)^{-1}d\lambda.$$

Proof. If (3.2) holds, then for every complex number λ, we have

$$(A - \lambda I)X - X(B - \lambda I) = C.$$

If $A - \lambda I$ and $B - \lambda I$ are invertible, this gives

$$X(B - \lambda I)^{-1} - (A - \lambda I)^{-1}X = (A - \lambda I)^{-1}C(B - \lambda I)^{-1}.$$

The theorem now follows by integrating over Γ and noting that

$$\frac{1}{2\pi i} \int_\Gamma (B - \lambda I)^{-1}d\lambda = 0 \text{ and } -\frac{1}{2\pi i} \int_\Gamma (A - \lambda I)^{-1}d\lambda = I.$$

Q. E. D.

Now we are going to establish a representation for the solutions of the Sylvester equation in terms of the operator exponential. Let

$$\alpha(A) := \sup \Re \, \sigma(A), \beta(A) := \inf \Re \, \sigma(A).$$

Theorem 2.4. *Suppose that*

$$\alpha(A) < \beta(B). \qquad (3.4)$$

Then the unique solution X of (3.1) is representable as

$$X = -\int_0^\infty e^{At}Ce^{-Bt}dt. \qquad (3.5)$$

Moreover, for any $\epsilon < \beta(B) - \alpha(A)$ there is a constant c_ϵ, such that

$$\|X\| \le \frac{c_\epsilon \|C\|}{\beta(B) - \alpha(A) - \epsilon}. \qquad (3.6)$$

Proof. Recall that

$$e^{At} = -\frac{1}{2\pi i} \int_\Gamma e^{\lambda t} R_\lambda(A) d\lambda \quad (t \geq 0),$$

where Γ is a contour surrounding $\sigma(A)$ with $\sup_{z \in \Gamma} \Re z = \alpha(A) + \epsilon/2$. Then

$$\|e^{At}\| \leq c_A e^{(\alpha(A) + \epsilon/2)t} \quad (t \geq 0, c_A = const).$$

Similarly,

$$\|e^{-Bt}\| \leq c_B e^{(\epsilon/2 - \beta(B))t} \quad (t \geq 0, c_B = const).$$

Hence it follows that

$$\|e^{At}\| \|e^{-Bt}\| \leq c_A c_B e^{(\alpha(A) - \beta(B) + \epsilon)t} \quad (t \geq 0),$$

and therefore for sufficiently small ϵ,

$$\int_0^\infty \|e^{At}\| \|e^{-Bt}\| dt \leq c_A c_B \frac{1}{\beta(B) - \alpha(A) - \epsilon} < \infty. \tag{3.7}$$

From (3.5) it follows

$$AX - XB = -\int_0^\infty (Ae^{At} Ce^{-Bt} - e^{At} CBe^{-Bt}) dt =$$

$$-\int_0^\infty \left(\frac{de^{At}}{dt} Ce^{-Bt} + e^{At} C \frac{de^{-Bt}}{dt}\right) dt =$$

$$-\int_0^\infty \frac{d}{dt} (e^{At} Ce^{-Bt}) dt = -e^{At} Ce^{-Bt}\big|_{t=0}^{t=\infty}.$$

In view of the convergence of the integral, $e^{At} Ce^{-Bt} \to 0$ as $t \to \infty$. Thus

$$-e^{At} Ce^{-Bt}\big|_{t=0}^{t=\infty} = C.$$

This and (3.7) prove the result. Q. E. D.

Furthermore, let

$$\alpha(A) < 0 \text{ and } \beta(B) > 0. \tag{3.8}$$

Put $b = \max\{\|A\|, \|B\|\}$. Take in the Rosenblum theorem

$$\Gamma = \Gamma_a = [-ia, ia] \cup L_a \quad (a > b),$$

where

$$L_a = \{z = ae^{it} \in \mathbf{C} : \pi/2 < t < 3\pi/2\}.$$

Then

$$X = \frac{1}{2\pi} \int_{-a}^{a} (A - iyI)^{-1} C (B - iyI)^{-1} dy + J_a \qquad (3.9)$$

where

$$J_a = \frac{a}{2\pi} \int_{\pi/2}^{3\pi/2} (A - ae^{it}I)^{-1} C (B - ae^{it}I)^{-1} e^{it} dt.$$

But

$$\|(A - ae^{it}I)^{-1}\| \leq 1/(a - b), \|(B - ae^{it}I)^{-1}\| \leq 1/(a - b).$$

So

$$\|J_a\| \leq \frac{a\|C\|}{2(a - b)^2} \to 0 \text{ as } a \to \infty.$$

Taking into account (3.9) we get our next result.

Corollary 2.1. *Let conditions (3.8) hold. Then the unique solution to (3.1) can be represented as*

$$X = \frac{1}{2\pi} \int_{-\infty}^{\infty} (A - iyI)^{-1} C (B - iyI)^{-1} dy.$$

If A^{-1} is invertible then the quasi-Sylvester equation (2.1) is equivalent to the equation $A^{-1}Y - YB = A^{-1}C$.

Note that $\sigma(A^{-1}) = \{1/\mu : \mu \in \sigma(A)\}$. Now the Rosenblum theorem implies our next result.

Corollary 2.2. *Let A be invertible and the condition*

$$\sigma(A^{-1}) \cap \sigma(B) = \emptyset,$$

hold and Γ a union of closed contours in the plane, with total winding numbers 1 around $\sigma(A^{-1})$ and 0 around $\sigma(B)$. Then the solution of equation (2.1) can be expressed as

$$Y = \frac{1}{2\pi i} \int_{\Gamma} (I - \lambda A)^{-1} C (B - \lambda I)^{-1} d\lambda.$$

2.4   Additional representations for solutions of the Sylvester equation

If A is invertible, then (3.1) can be written as $X - A^{-1}XB = A^{-1}C$. So (3.1) takes the form (2.1) with A^{-1} instead of A and $A^{-1}C$ instead of C. Furthermore, let $r_{\text{low}}(A)$ denote the lower spectral radius of A: $r_{\text{low}}(A) = \inf\{|s| : s \in \sigma(A)\}$. The spectrum mapping theorem yields the relation

$$r_s(A^{-1}) = \frac{1}{r_{\text{low}}(A)}.$$

According to (2.3) assume that

$$r_s(B) < r_{\text{low}}(A). \tag{4.1}$$

Then Lemma 2.1 implies

Corollary 2.3. *Let condition (4.1) hold. Then the unique solution X of (3.1) is representable as*

$$X = \sum_{k=0}^{\infty} A^{-k-1}CB^k. \tag{4.2}$$

We have

$$\|A^{-k}\| \le \frac{c_A}{(r_{\text{low}}(A) - \epsilon)^k} \quad (k = 1, 2, ...; \ c_A = const)$$

for an arbitrary $0 < \epsilon < r_{\text{low}}(A)$. So,

$$\|A^{-k-1}\|\|B^k\| \le c_A c_B \frac{(r_s(B) + \epsilon)^k}{(r_{\text{low}}(A) - \epsilon)^{k+1}}. \tag{4.3}$$

If

$$2\epsilon < r_{\text{low}}(A) - r_s(B), \tag{4.4}$$

then

$$\frac{r_s(B) + \epsilon}{r_{\text{low}}(A) - \epsilon} < 1.$$

Therefore condition (4.1) holds and (4.2) is valid. Besides,

$$\|X\| \le \|C\| \sum_{k=0}^{\infty} \|A^{-k-1}\|\|B^k\| \le c_A c_B \|C\| \sum_{k=0}^{\infty} \frac{(r_s(B) + \epsilon)^k}{(r_{\text{low}}(A) - \epsilon)^{k+1}} =$$

$$\frac{c_A c_B \|C\|}{r_{\text{low}}(A) - \epsilon} \left(1 - \frac{r_s(B) + \epsilon}{r_{\text{low}}(A) - \epsilon}\right)^{-1}.$$

We thus obtain

$$\|X\| \le \frac{c_A c_B \|C\|}{r_{\text{low}}(A) - r_s(B) - 2\epsilon},$$ (4.5)

provided (4.4) holds.

Furthermore, since

$$(Ie^{-i\omega} - aA^{-1})^{-1}A^{-1} = (Ae^{-i\omega} - aI)^{-1},$$

reducing (3.1) to the equation $X - A^{-1}XB = A^{-1}C$ and making use of Lemma 2.2, we get

Lemma 2.3. *Let there be an $a \in \mathbf{C}, a \ne 0$, such that $|a| < r_{\text{low}}(A)$ and $r_s(B) < |a|$. Then the unique solution X to (3.1) is representable by the integral*

$$X = \frac{a}{2\pi} \int_0^{2\pi} (Ae^{-i\omega} - aI)^{-1}C(ae^{i\omega}I - B)^{-1}d\omega.$$

Note that in the paper [8] the equation

$$\sum_{k=0}^m A^{m-k}XB^k = C$$ (4.6)

is investigated. Besides, the following result have been established.

Theorem 2.5. *If the spectra of A and B are in the sector*

$$\{z \in \mathbf{C} : z \ne 0, -\pi/(m+1) < \arg z < \pi/(m+1)\}$$

then (4.6) has a unique solution X_m representable as

$$X_m = \frac{\sin\left(\pi/(m+1)\right)}{\pi} \int_0^\infty (tI + A^{m+1})^{-1}C(tI + B^{m+1})^{-1}t^{1/(m+1)}dt.$$

Specialising the latter equality to the case $m = 1$ we obtain yet another formula for the solution of the Sylvester equation.

Corollary 2.4. *If the spectrum of A is contained in the open right half plane and that of B in the open left half plane, then the solution X of (3.1) can be represented as*

$$X = \frac{1}{\pi} \int_0^\infty (tI + A^2)^{-1}C(tI + B^2)^{-1}t^{1/2}dt.$$

2.5   Polynomial operator equations

In this section we consider the equations

$$\sum_{k=0}^{m} c_{m-k} A^k X B^k = C \tag{5.1}$$

and

$$\sum_{k=0}^{m} c_{m-k} A^k X B^{m-k} = C, \tag{5.2}$$

where $c_k \in \mathbf{C}$ $(k = 1, ..., m)$, $c_0 = 1$, $A \in \mathcal{B}(\mathcal{X})$, $B \in \mathcal{B}(\mathcal{Y})$ and $C \in B(\mathcal{Y}, \mathcal{X})$.

Theorem 2.6. *Let x_k $(k = 1, ..., m)$ be the roots taken with the multiplicities of the polynomial*

$$p(x) := \sum_{k=0}^{m} c_k x^{m-k}$$

and let

$$r_s(A) r_s(B) < \min_k |x_k|. \tag{5.3}$$

Then (5.1) has a unique solution X which can be represented by

$$X = (-1)^m \sum_{j_1, j_2, ..., j_m = 0}^{\infty} \frac{1}{x_1^{j_1+1} x_2^{j_2+1} \ ... \ x_m^{j_m+1}} A^{j_1+j_2+...+j_m} C B^{j_1+j_2+...+j_m},$$

and the series converges in the operator norm.

The proof of this theorem is presented in the next section.

Now consider equation (5.2), assuming that B is invertible; from that equation we have

$$\sum_{k=0}^{m} c_{m-k} A^k X B^{-k} = C B^{-m}. \tag{5.4}$$

Recall that $r_{\text{low}}(B) = \inf_{\lambda \in \sigma(B)} |\lambda|$. Taking into account that $r_{\text{low}}(B) = 1/r_s(B^{-1})$ and applying Theorem 2.6 to (5.4) with B^{-1} instead of B, we get

Corollary 2.5. *Let*

$$r_s(A) < r_{\text{low}}(B) \min_k |x_k|.$$

Then (5.2) has a unique solution X which can be represented by

$$X = (-1)^m \sum_{j_1, j_2, ..., j_m = 0}^{\infty} \frac{1}{x_1^{j_1+1} x_2^{j_2+1} \ ... \ x_m^{j_m+1}} A^{j_1+j_2+...+j_m} C B^{-m-j_1-j_2-...-j_m}$$

and the series converges in the operator norm.

2.6   Proof of Theorem 2.6

Again use the operators A_l and B_r defined by $A_l X := AX$ and $B_r X := XB$, respectively. Recall that A_l and B_r commute.

Introduce the operator-valued function

$$\Phi(p(zw), A, B) = -\frac{1}{4\pi^2} \int_{L_B} \int_{L_A} p(zw) R_z(A_l) R_w(B_r) dw \, dz, \qquad (6.1)$$

where L_A, L_B are closed Jordan contours surrounding $\sigma(A)$ and $\sigma(B)$, respectively. So for any $C \in \mathcal{B}(\mathcal{Y}, \mathcal{X})$,

$$\Phi(p(zw), A, B)C = -\frac{1}{4\pi^2} \int_{L_B} \int_{L_A} p(zw) R_z(A) C R_w(B) dw \, dz.$$

If

$$p(zw) \neq 0 \quad (z \in \sigma(A), w \in \sigma(B)), \qquad (6.2)$$

then due to Theorem 2.2 equation (5.1) has a unique solution which can be written as

$$X = \Phi(1/p(zw), A, B)C.$$

Since

$$p(x) = \prod_{k=1}^{m} (x - x_k),$$

making use of the property c) from Section 2.1, we can rewrite (6.1) as

$$\Phi(zw - x_1, A, B)\Phi(zw - x_2, A, B) \cdots \Phi(zw - x_m, A, B)X = C.$$

But $\Phi(zw - x_k, A, B) = A_l B_r - x_k I$. Thus,

$$\prod_{k=1}^{m} (A_l B_r - x_k I)X = C$$

and therefore, we get

Lemma 2.4. *Under condition (6.2) equation (5.1) has a unique solution X defined by*

$$X = \prod_{k=1}^{m} (A_l B_r - x_k I)^{-1} C. \qquad (6.3)$$

Furthermore, for any $a \in \mathbf{C}$, satisfying

$$|a| > r_s(A)r_s(B), \tag{6.4}$$

consider the operator $Z(a)$ defined by

$$Z(a)C = -\sum_{k=0}^{\infty} \frac{1}{a^{k+1}} A_l^k B_r^k C = -\sum_{k=0}^{\infty} \frac{1}{a^{k+1}} A^k C B^k. \tag{6.5}$$

The series converges in the operator norm and

$$A_l B_r Z(a) = AZ(a)B = -\sum_{k=0}^{\infty} \frac{1}{a^{k+1}} A^{k+1} C B^{k+1} =$$

$$-\sum_{k=0}^{\infty} \frac{1}{a^k} A^k C B^k + C = aZ(a) + C.$$

We thus obtain the following result.

Lemma 2.5. *Let condition (6.4) hold. Then operator $Z(a)$ defined by (6.5) is the unique solution to the equation $AZ(a)B - aZ(a) = C$ and $Z(a) = (A_l B_r - aI)^{-1} C$.*

Proof of Theorem 2.6: Put

$$Y_1 = (A_l B_r - x_1 I)^{-1} C, Y_2 = (A_l B_r - x_2 I)^{-1} Y_1, ..., Y_k = (A_l B_r - x_k I)^{-1} Y_{k-1}.$$

Lemma 2.4 implies $X = Y_m$. Due to the previous lemma Y_k are solutions to the equations

$$AY_k B - x_k Y_k = Y_{k-1} \ (k = 2, ..., m), \ \ AY_1 B - x_1 Y_1 = C$$

and

$$Y_j = -\sum_{k=0}^{\infty} \frac{1}{x_j^{k+1}} A^k Y_{k-1} B^k,$$

provided (5.8) holds. So

$$Y_1 = -\sum_{k=0}^{\infty} \frac{1}{x_1^{k+1}} A^k C B^k, Y_2 = -\sum_{j=0}^{\infty} \frac{1}{x_2^{j+1}} A^j Y_1 B^j =$$

$$\sum_{j=0}^{\infty} \frac{1}{x_2^{j+1}} A^j \sum_{k=0}^{\infty} \frac{1}{x_1^{k+1}} A^k C B^k B^j = \sum_{j,k=0}^{\infty} \frac{1}{x_1^{k+1} x_2^{j+1}} A^{k+j} C B^{k+j}.$$

Continuing this process for $j = 3, ..., m$, according to Lemma 2.4, we prove the theorem. Q. E. D.

2.7 Additional representations of solutions to equation (5.1)

Consider the equation

$$x_k Y - AYB = C \quad (k = 1, ..., m). \tag{7.1}$$

If $r_s(A)r_s(B) < x_k$, then there is a $d_k \in \mathbf{C}, d_k \neq 0$, such that

$$|d_k|r_s(A) < |x_k| \text{ and } r_s(B) < |d_k| \quad (k = 1, ...m). \tag{7.2}$$

Rewrite (7.1) as

$$Y - \frac{1}{x_k}AYB = \frac{1}{x_k}C.$$

Due to Lemma 2.2 we have

$$Y = -(A_l B_r - x_k)^{-1}C = \hat{Z}(d_k, x_k)C,$$

where

$$\hat{Z}(d_k, x_k)C = -\frac{d_k}{2\pi} \int_0^{2\pi} (Ix_k e^{-i\omega} - d_k A)^{-1}C(d_k e^{i\omega}I - B)^{-1}d\omega.$$

Now (6.3) implies

Theorem 2.7. *Let* x_k $(k = 1, ..., m)$ *be the roots of* $p(x)$ *and conditions (7.2) hold. Then (5.1) has a unique solution* X, *which can be represented as*

$$X = \prod_{k=1}^m \hat{Z}(d_k, x_k)C.$$

2.8 Additional representations of solutions to equation (5.2)

Corollary 2.5 does not enable us to consider, many operator equations, for example, the Lyapunov equation, since $r_s(A) = r_s(A^*)$. Because of this we are going to derive the representation of solutions to (5.2) under other conditions. To this end put

$$\hat{p}(z, w) = \sum_{k=0}^m c_{m-k}z^k w^{m-k} = w^m p(z/w) =$$

$$w^m \prod_{k=1}^m (z/w - x_k) = \prod_{k=1}^m (z - x_k w).$$

So due to the property c) from (2.1), equation (5.2) can be written as

$$\prod_{k=1}^{m}(A_l - x_k B_r)X = C$$

and therefore

$$X = \prod_{k=1}^{m}(A_l - x_k B_r)^{-1}C, \tag{8.1}$$

provided

$$\hat{p}(z,w) \neq 0 \quad (z \in \sigma(A), z \in \sigma(B)). \tag{8.2}$$

From Lemma 2.2, it directly follows that the equation

$$AX - aXB = C \quad (a \in \mathbf{C}) \tag{8.3}$$

has a unique solution X_a, which can be represented as

$$X_a = -\int_0^\infty e^{At}Ce^{-aBt}dt, \tag{8.4}$$

provided

$$\beta(aB) > \alpha(A). \tag{8.5}$$

A solution of (8.3) is also given by $X_a = (A_l - aB_r)^{-1}C$. So under condition (8.5) we have

$$(A_l - aB_r)^{-1}C = -\int_0^\infty e^{At}Ce^{-aBt}dt.$$

Now assume that

$$\beta(x_k B) > \max_{k=1,\dots,m} \alpha(A). \tag{8.6}$$

Then

$$(A_l - x_k B_r)^{-1}C = -\int_0^\infty e^{At}Ce^{-x_k Bt}dt.$$

Put

$$W_1 = (A_l - x_1 B_r)^{-1}C, W_2 = (A_l - x_2 B_r)^{-1}Y_1,$$

$$\dots, W_k = (A_l - x_k B_r)^{-1}W_{k-1}.$$

Then $X = W_m$. Due to the previous lemma Y_k are solutions to the equations

$$AW_k - x_k B = W_{k-1} \quad (k = 2, \dots, m), \quad AY_1 B - x - 1Y_1 = C.$$

and

$$W_j = -\int_0^\infty e^{At} W_{j-1} e^{-x_k Bt} dt$$

provided (8.6) holds. So

$$W_1 = \int_0^\infty e^{At_1} C e^{-x_1 Bt_1} dt_1, \quad W_2 = \int_0^\infty e^{-At_2} W_1 e^{x_2 Bt_2} dt_2$$

$$= \int_0^\infty \int_0^\infty e^{(t_1+t_2)A} C e^{-(x_2 t_2 + x_1 t_1)B} dt_1 dt_2.$$

Continuing this process for $j = 3, ..., m$, according to (8.2) we obtain

$$X = (-1)^m \int_0^\infty ... \int_0^\infty e^{(t_1+...+t_m)A} C e^{-(x_1 t_1 + ... + x_m t_m)B} dt_1 ... dt_m. \quad (8.7)$$

We thus arrive at

Theorem 2.8. *Let x_k ($k = 1, ..., m$) be the roots of $p(x)$ and condition (8.6) hold. Then (5.2) has a unique solution X, which can be represented by (8.7).*

2.9 Equations with scalar type spectral operators

Let Σ be the algebra of the Borel sets of $\mathbf{C}$.

By a *countably additive spectral measure* in a complex Banach space $\mathcal{X}$ we mean a map E that assigns to every Borel set of the complex plane a projection in $\mathcal{B}(\mathcal{X})$, such that

$$E(\mathbf{C}) = I, E(\emptyset) = 0, E(\omega_1 \cap \omega_2) = E(\omega_1)E(\omega_2) \quad (\omega_1, \omega_2 \in \Sigma)$$

and, in addition,

$$E(\cup_{k=1}^\infty \omega_k) = \sum_{k=1}^\infty E(\omega_k) \text{ whenever } \omega_j \cap \omega_k = \emptyset \quad (j \neq k, \ \omega_k \in \Sigma).$$

It is supposed that the right-hand series converge in the strong operator topology. Every countably additive spectral measure is bounded:

$$\sup_{\omega \in \Sigma} \|E(\omega)\| < \infty.$$

A countably additive spectral measure E_A is said to be *the resolution of the identity* for an operator $A \in \mathcal{B}(\mathcal{X})$, if the following condition is satisfied:

$$E_A(\omega)A = AE_A(\omega) \text{ for all } \omega \in \Sigma.$$

An operator $A \in \mathcal{B}(\mathcal{X})$ is said to be *a scalar type spectral operator, or simply a scalar type operator*, if

$$A = \int_{\mathbf{C}} \lambda E_A(d\lambda).$$

As is shown in [19], for any scalar type operator A and any function f defined on $\mathbf{C}$, E-measurable and bounded on $\sigma(A)$, one can define the operator function

$$f(A) = \int_{\mathbf{C}} f(\lambda) E_A(d\lambda).$$

Besides, there is a constant c_A, independent on f, such that

$$\|f(A)\| \le c_A \sup_{\lambda \in \sigma(A)} |f(\lambda)|.$$

In particular,

$$\|e^{tA}\| \le c_A e^{t\alpha(A)} \ (t \ge 0) \text{ and } \|A^k\| \le c_A r_s^k(A) \ (k = 1, 2, ...).$$

Again put

$$P(\lambda, \mu) = \sum_{j,k=1}^{m} c_{jk} \lambda^j \mu^k.$$

Theorem 2.9. *Let $A \in \mathcal{B}(\mathcal{X}), B \in \mathcal{B}(\mathcal{Y})$ be scalar type spectral operators, and condition (1.8) hold. Then the (unique) solution X to the equation*

$$\sum_{j,k=0}^{m} c_{jk} A^j X B^k = C \ (C \in \mathcal{B}(\mathcal{Y}, \mathcal{X}), c_{jk} \in \mathbf{C}, j, k = 1, ..., m) \qquad (9.1)$$

can be expressed as

$$X = \int_{\mathbf{C}} \int_{\mathbf{C}} \frac{1}{P(\lambda, \mu)} E_A(d\lambda) C E_B(d\mu), \qquad (9.2)$$

where E_A and E_B are the resolutions the identity of A and B, respectively.

Proof. Indeed, from (9.2) we have

$$\sum_{j,k=0}^{m} c_{jk} A^j X B^k =$$

$$\sum_{j,k=0}^{m} c_{jk} \int_{\mathbf{C}} \int_{\mathbf{C}} \frac{1}{P(\lambda, \mu)} A^j E_A(d\lambda) C E_B(d\mu) B^k =$$

$$\sum_{j,k=0}^{m} c_{jk} \int_{\mathbf{C}} \int_{\mathbf{C}} \frac{1}{P(\lambda, \mu)} \lambda^j E_A(d\lambda) C \mu^k E_B(d\mu) =$$

$$\int_{\mathbf{C}} \int_{\mathbf{C}} E_A(d\lambda) C E_B(d\mu) = C,$$

as claimed. Q. E. D.

2.10 Perturbations of two-sided Sylvester equations

Consider the two-sided Sylvester equations

$$\sum_{s=1}^{m} A_{1s} X A_{2s} = C \tag{10.1}$$

and

$$\sum_{s=1}^{m} \tilde{A}_{1s} \tilde{X} \tilde{A}_{2s} = C \tag{10.2}$$

$(A_{1s}, \tilde{A}_{1s} \in \mathcal{B}(\mathcal{X}); A_{2s}, \tilde{A}_{2s} \in \mathcal{B}(\mathcal{Y}); \ C \in \mathcal{B}(\mathcal{Y}, \mathcal{X}), s = 1, ..., m)$ and denote

$$\Delta_{ls} = \tilde{A}_{ls} - A_{ls} \text{ and } Y = \tilde{X} - X.$$

Subtracting (10.1) from (10.2), we have

$$\sum_{s=1}^{m} \tilde{A}_{1s} Y \tilde{A}_{2s} + \Delta_{1s} X \tilde{A}_{2s} + A_{1s} X \Delta_{2s} = 0.$$

Hence,

$$\sum_{s=1}^{m} \tilde{A}_{1s} Y \tilde{A}_{2s} = -\Phi,$$

where

$$\Phi := \sum_{s=1}^{m} (\Delta_{1s} X \tilde{A}_{2s} + A_{1s} X \Delta_{2s}).$$

Assume that the operators $\tilde{Z}$ and Z defined by

$$ZC = \sum_{s=1}^{m} A_{1s} C A_{2s}$$

and

$$\tilde{Z}C = \sum_{s=1}^{m} \tilde{A}_{1s} C \tilde{A}_{2s}$$

be boundedly invertible. Then $Y = \tilde{Z}^{-1}\Phi$ and

$$\|Y\| \le \|\tilde{Z}^{-1}\| \|\Phi\| \le$$

$$\|\tilde{Z}^{-1}\| \|X\| \sum_{s=1}^{m} (\|\Delta_{1s}\| \|\tilde{A}_{2s}\| + \|A_{1s}\| \|\Delta_{2s}\|).$$

Hence, we get

Theorem 2.10. *Let Z and $\tilde{Z}$ be invertible. Then equations (10.1) and (10.2) have unique solutions X and $\tilde{X}$, respectively. Moreover,*

$$\|X - \tilde{X}\| \le \|C\| \|\tilde{Z}^{-1}\| \|Z^{-1}\| \sum_{s=1}^{m} (\|\Delta_{1s}\| \|\tilde{A}_{2s}\| + \|A_{1s}\| \|\Delta_{2s}\|). \tag{10.3}$$

Note that

$$(\tilde{Z} - Z)C = \sum_{s=1}^{m}(\tilde{A}_{1s}C\tilde{A}_{2s} - A_{1s}CA_{2s}) = \sum_{s=1}^{m}(\Delta_{1s}C\tilde{A}_{2s} + A_{1s}C\Delta_{2s})$$

and therefore, $\|\tilde{Z} - Z\| \leq \delta$, where

$$\delta := \sum_{s=1}^{m}(\|\Delta_{1s}\|\|\tilde{A}_{2s}\| + \|A_{1s}\|\|\Delta_{2s}\|).$$

Now (10.3) implies.   ∎

Corollary 2.6. *Let Z be invertible and $\delta\|Z^{-1}\| < 1$. Then $\tilde{Z}$ is also invertible, equations (10.1) and (10.2) have unique solutions X and $\tilde{X}$, respectively. Moreover,*

$$\|\tilde{Z}^{-1}\| \leq \|Z^{-1}\|(1 - \delta\|Z^{-1}\|)^{-1},$$

and

$$\|X - \tilde{X}\| \leq \|C\|\|Z^{-1}\|^{2}(1 - \delta\|Z^{-1}\|)^{-1}\sum_{s=1}^{m}(\|\Delta_{1s}\|\|\tilde{A}_{2s}\| + \|A_{1s}\|\|\Delta_{2s}\|).$$

2.11   Differentiating of solutions to two-sided Sylvester equations

Let $[a, b]$ be a finite or infinite real segment. A function $Z : [a, b] \to \mathcal{B}(\mathcal{Y}, \mathcal{X})$ is continuous at a point $t_0 \in [a, b]$ if

$$\|Z(t) - Z(t_0)\| \to 0 \text{ as } t \to t_0 \quad (\|Z(t)\| = \|Z(t)\|_{\mathcal{B}(\mathcal{Y}, \mathcal{X})}).$$

$Z(t)$ is continuous if it is continuous at each point of $[a, b]$. The norm of a continuous function is a continuous function.

Let $C([a, b]; \mathcal{B}(\mathcal{Y}, \mathcal{X}))$ be the Banach space of all bounded continuous functions acting from $[a, b]$ into $\mathcal{B}(\mathcal{Y}, \mathcal{X})$ with the norm

$$\|Z\|_{C([a,b];\mathcal{B}(\mathcal{Y},\mathcal{X}))} = \sup_{a \leq t \leq b} \|Z(t)\|.$$

We will say that $Z(t) \in C([a, b]; \mathcal{B}(\mathcal{Y}, \mathcal{X}))$ has at a point t_0 a right (left) derivative if there exists $M \in \mathcal{B}(\mathcal{Y}, \mathcal{X})$ such that

$$\left\|\frac{Z(t_0 + \delta t) - Z(t_0)}{\delta t} - M\right\| \to 0$$

as $\delta t \to +0$ ($\delta t \to -0$). One then writes

$$\frac{d_+ Z(t_0)}{dt} = M \quad (\frac{d_- Z(t_0)}{dt} = M).$$

If the left and the right derivatives exist, Z is continuous and

$$\frac{d_- Z(t_0)}{dt} = \frac{d_+ Z(t_0)}{dt},$$

then $Z(t)$ is differentiable at the point t_0 and its derivative is

$$Z'(t_0) = \frac{dZ(t_0)}{dt} = M.$$

The function $Z(t)$ is differentiable on $[a, b]$ if it is differentiable at each point of (a, b), has the right derivative at a, and has the left derivative at b.

Furthermore, let $A_{1s}(t) \in \mathcal{B}(\mathcal{X}), A_{2s}(t) \in \mathcal{B}(\mathcal{Y})$ $(s = 1, ..., m)$ be operators defined and differentiable on $[a, b]$. Consider the equation

$$\sum_{s=1}^{m} A_{1s}(t) X(t) A_{2s}(t) = C, \tag{11.1}$$

where $C \in \mathcal{B}(\mathcal{Y}, \mathcal{X})$ is independent of t. For any $S \in \mathcal{B}(\mathcal{Y}, \mathcal{X})$ define the operator $Z(t)$ by

$$Z(t)S := \sum_{s=1}^{m} A_{1s}(t) S A_{2s}(t). \tag{11.2}$$

Assume that $Z(t)$ has on $[a, b]$ a bounded inverse $Z^{-1}(t) \in \mathcal{B}(\mathcal{Y}, \mathcal{X})$. Then for each $t \in [a, b]$, (11.1) has a unique solution $X(t) = Z^{-1}(t)C$. Differentiating (11.1), we obtain

$$\sum_{s=1}^{m} A_{1s}(t) X'(t) A_{2s}(t) = \Psi(t), \tag{11.3}$$

where

$$\Psi(t) = - \sum_{s=1}^{m} (A'_{1s}(t) X(t) A_{2s}(t) + A_{1s}(t) X(t) A'_{2s}(t)).$$

Hence, $X'(t) = Z^{-1}(t)\Psi(t)$. Take into account that

$$\|\Psi(t)\| \le \|X(t)\| \sum_{s=1}^{m} (\|A'_{1s}(t)\| \|A_{2s}(t)\| + \|A_{1s}(t)\| \|A'_{2s}(t)\|).$$

We thus have proved the following result.

Theorem 2.11. *Let $A_{1s}(t), A_{2s}(t)$ $(s = 1, ..., m)$ have on $[a, b]$ bounded derivatives. If, in addition, $Z(t)$ is boundedly invertible on $[a, b]$, then (11.1) has for each $t \in [a, b]$ a unique differentiable solution and*

$$\|X'(t)\| \le \|Z^{-1}(t)\|^2 \|C\| \sum_{s=1}^{m} (\|A'_{1s}(t)\| \|A_{2s}(t)\| + \|A_{1s}(t)\| \|A'_{2s}(t)\|).$$

2.12　Perturbed Sylvester equations and spectral variations

Again $A, \tilde{A} \in \mathcal{B}(\mathcal{X})$ and $B, \tilde{B} \in \mathcal{B}(\mathcal{Y})$. In addition, $C, \tilde{C} \in \mathcal{B}(\mathcal{Y}, \mathcal{X})$. Consider the Sylvester equations

$$AX - XB = C \tag{12.1}$$

and

$$\tilde{A}\tilde{X} - \tilde{X}\tilde{B} = \tilde{C}. \tag{12.2}$$

Due to (3.2) the necessary and sufficient condition for the existence of the unique solution to (12.2) is

$$\tilde{\lambda} - \tilde{\mu} \neq 0 \quad (\tilde{\lambda} \in \sigma(\tilde{A}), \tilde{\mu} \in \sigma(\tilde{B})). \tag{12.3}$$

But for any $\tilde{\lambda}$ and $\tilde{\mu}$ there are $\lambda \in \sigma(A)$ and $\mu \in \sigma(B)$, such that

$$|\lambda - \tilde{\lambda}| \leq sv_A(\tilde{A}) \text{ and } |\tilde{\mu} - \mu| \leq sv_B(\tilde{B}), \tag{12.4}$$

where $sv_A(\tilde{A})$ is the spectral variation of $\tilde{A}$ with respect to A. Consequently, if

$$\inf_{\lambda \in \sigma(A), \mu \in \sigma(B)} |\lambda - \mu| > sv_A(\tilde{A}) + sv_B(\tilde{B}), \tag{12.5}$$

then condition (12.3) holds. So we arrive at the following lemma.

Lemma 2.6. *Let condition (12.5) hold. Then (12.2) has a unique solution.*

This lemma enables us to apply the estimates for the spectral variations established below to perturbed equations.

Now consider the equations

$$X - AXB = C \tag{12.6}$$

and

$$\tilde{X} - \tilde{A}\tilde{X}\tilde{B} = \tilde{C}. \tag{12.7}$$

Due to Theorem 2.2 the necessary and sufficient condition for the existence of the unique solution to (12.7) is

$$\tilde{\lambda}\tilde{\mu} \neq 1 \quad (\tilde{\lambda} \in \sigma(\tilde{A}), \tilde{\mu} \in \sigma(\tilde{B})). \tag{12.8}$$

Taking into account (12.4) we obtain our next result.

Lemma 2.7. *Let the condition*

$$(r_s(A) + sv_A(\tilde{A}))(r_s(B) + sv_B(\tilde{B})) < 1$$

hold. Then equation (12.7) has a unique solution.

2.13　Comments to Chapter 2

Sections 2.1-2.3 are based on [14, Chapter I], Theorems 2.6, 2.7 and 2.8 appear in [44]. Theorems 2.9, 2.10 and 2.11 are probably new.

Chapter 3

Functions of Finite Matrices

This chapter is devoted to functions of finite dimensional operators. In particular, we establish norm estimates for the resolvents and functions, regular on the convex hull $co(A)$ of the spectrum. We also estimate the functions of the form $\frac{1}{f}(A)$, where f is regular on $co(A)$. By these estimates in the next chapter we investigate matrix equations. In addition, spectrum perturbations of matrices are explored.

3.1 Departure from normality

Let $\mathbf{C}^n$ be the n-dimensional complex Euclidean space with a scalar product $(.,.)$ and the norm $\|.\| = \sqrt{(.,.)}$; $\mathbf{C}^{n \times n}$ is the set of $n \times n$-matrices; $\|A\|$ denotes its operator (spectral) norm of $A \in \mathbf{C}^{n \times n}$ and

$$N_p(A) = (\text{trace}\,(AA^*)^{p/2})^{1/p} \quad (1 \le p < \infty).$$

For an $n \times n$-matrix A introduce the quantity (the departure from normality)

$$g(A) = (N_2^2(A) - \sum_{k=1}^{n} |\lambda_k(A)|^2)^{1/2}, \qquad (1.1)$$

where $\lambda_k(A)$ are the eigenvalues of A taken with their multiplicities. Since

$$\sum_{k=1}^{n} |\lambda_k(A)|^2 \ge |\text{trace}\,(A^2)|,$$

we get

$$g^2(A) \le N_2^2(A) - |\text{trace}\,(A^2)|. \qquad (1.2)$$

Example 3.1. Consider the matrix

$$A = \begin{pmatrix} a_{11} & a_{12} \\ a_{21} & a_{22} \end{pmatrix}$$

where a_{jk} $(j, k = 1, 2)$ are real numbers, assuming that the eigenvalues of A are real: $(\text{trace } A)^2 > 4 \det A$. Then

$$|\lambda_1(A)|^2 + |\lambda_2(A)|^2 = \text{trace } (A^2).$$

But trace $(A^2) = a_{11}^2 + 2a_{12}a_{21} + a_{22}^2$. Consequently,

$$g^2(A) = N_2^2(A) - |\lambda_1(A)|^2 - |\lambda_2(A)|^2 =$$

$$a_{11}^2 + a_{12}^2 + a_{21}^2 + a_{22}^2 - (a_{11}^2 + 2a_{12}a_{21} + a_{22}^2).$$

Hence,

$$g(A) = |a_{12} - a_{21}|. \tag{1.3}$$

Recall that by Schur's theorem [68, Section I.4.10.2], there is an orthogonal normal *(Schur's) basis* $\{e_k\}_{k=1}^n$, in which A has the triangular representation

$$Ae_k = \sum_{j=1}^{k} a_{jk} e_j \text{ with } a_{jk} = (Ae_k, e_j) \ (k = 1, ..., n).$$

Schur's basis is not unique. We can write

$$A = D + V \ (\sigma(A) = \sigma(D)) \tag{1.4}$$

with a normal (diagonal) operator D defined by

$$De_j = \lambda_j(A)e_j \ (j = 1, ..., n)$$

and a nilpotent operator V defined by

$$Ve_k = \sum_{j=1}^{k-1} a_{jk} e_j \ (k = 2, ..., n), \ Ve_1 = 0.$$

Equality (1.4) is called *the triangular representation of A; D and V are called the diagonal part and nilpotent part of A*, respectively. Put

$$P_j = \sum_{k=1}^{j} (., e_k) e_k \ (j = 1, ..., n), \ P_0 = 0.$$

$\{P_k\}_{k=1}^n$ is called *the maximal chain of the invariant projections of A*. It has the properties

$$0 = P_0 \mathbf{C}^n \subset P_1 \mathbf{C}^n \subset ... \subset P_n \mathbf{C}^n = \mathbf{C}^n$$

with dim $(P_k - P_{k-1})\mathbf{C}^n = 1$ and

$$AP_k = P_k A P_k; \quad VP_k = P_{k-1}VP_k; \quad DP_k = DP_k \quad (k = 1, ..., n).$$

So A, V and D have the joint invariant subspaces. We can write

$$D = \sum_{k=1}^{n} \lambda_k(A)\Delta P_k,$$

where $\Delta P_k = P_k - P_{k-1} \quad (k = 1, ..., n)$.

Lemma 3.1. *One has $N_2(V) = g(A)$, where V is the nilpotent part of A.*

Proof. Let D be the diagonal part of A. Then, both matrices V^*D and D^*V are nilpotent. Therefore,

$$\text{trace } (D^*V) = 0. \tag{1.5}$$

It is easy to see that

$$\text{trace } (D^*D) = \sum_{k=1}^{n} |\lambda_k(A)|^2. \tag{1.6}$$

From (1.4) we obtain

$$N_2^2(A) = \text{trace } (D + V)^*(V + D) = \text{trace } (V^*V + D^*D) =$$

$$N_2^2(V) + \sum_{k=1}^{n} |\lambda_k(A)|^2,$$

and the required equality is proved. Q. E. D.

Lemma 3.2. *For any linear operator A in $\mathbf{C}^n$,*

$$N_2^2(V) = 2N_2^2(A_I) - 2\sum_{k=1}^{n} |\Im \, \lambda_k(A)|^2,$$

where V is the nilpotent part of A and $A_I := \Im A = (A - A^*)/2i$.

Proof. Clearly,

$$-4(A_I)^2 = (A - A^*)^2 = AA - AA^* - A^*A + A^*A^*.$$

But due to (1.4) and (1.5)

$$\text{trace } (A - A^*)^2 = \text{trace } (V + D - V^* - D^*)^2 =$$

$$\text{trace } [(V - V^*)^2 + (V - V^*)(D - D^*) +$$

$$(D - D^*)(V - V^*) + (D - D^*)^2] =$$

$$\text{trace } (V - V^*)^2 + \text{trace } (D - D^*)^2.$$

Hence,

$$N_2^2(A_I) = N_2^2(V_I) + N_2^2(D_I),$$

where

$$V_I = (V - V^*)/2i \text{ and } D_I = (D - D^*)/2i.$$

It is not hard to see that

$$N_2^2(V_I) = \frac{1}{2} \sum_{m=1}^{n} \sum_{k=1}^{m-1} |a_{km}|^2 = \frac{1}{2} N_2^2(V),$$

where a_{jk} are the entries of V in the Schur basis. Consequently,

$$N_2^2(V) = 2N^2(A_I) - 2N_2^2(D_I).$$

But

$$N_2^2(D_I) = \sum_{k=1}^{n} |\Im \lambda_k(A)|^2.$$

We thus, arrive at the required equality. Q. E. D.

From Lemmas 3.1 and 3.2 we obtain

Theorem 3.1. *Let $A \in \mathbf{C}^{n \times n}$. Then*

$$g^2(A) = 2N_2^2(A_I) - 2\sum_{k=1}^{n} |\Im \lambda_k(A)|^2.$$

From this theorem one has $g^2(A) \leq 2N_2^2(A_I)$. Furthermore, we take into account that the nilpotent parts of the matrices A and $Ae^{i\tau} + zI$ with a real number τ and a complex one z, coincide. Hence, due to Lemma 3.1 we obtain the following

Corollary 3.1. *For any linear operator A in $\mathbf{C}^n$, a real number τ and a complex one z, the relation $g(e^{i\tau}A + zI) = g(A)$ holds.*

Corollary 3.2. *For arbitrary linear operators A, B in $\mathbf{C}^n$, having a joint Schur basis,*

$$g(A + B) \leq g(A) + g(B).$$

In fact, we have $V_{A+B} = V_A + V_B$, where V_{A+B}, V_A and V_B are the nilpotent parts of $A + B, A$ and B, respectively. Due to Lemma 3.1 the relations

$$g(A) = N_2(V_A), \; g(B) = N_2(V_B), \; g(A + B) = N_2(V_{A+B})$$

are true. Now the property of the norm implies the required result.

Corollary 3.3. *For any $n \times n$ matrix A and real numbers t, τ the equality*

$$N_2^2(Ae^{it} - A^* e^{-it}) - \sum_{k=1}^{n} |e^{it} \lambda_k(A) - e^{-it} \overline{\lambda}_k(A)|^2 =$$

$$N_2^2(Ae^{i\tau} - A^* e^{-i\tau}) - \sum_{k=1}^{n} |e^{i\tau} \lambda_k(A) - e^{-i\tau} \overline{\lambda}_k(A)|^2$$

is true.

The proof consists in replacing A by Ae^{it} and $Ae^{i\tau}$ and applying Theorem 3.1. In particular, take $t = 0$ and $\tau = \pi/2$. Due to Corollary 3.3,

$$N_2^2(A_I) - \sum_{k=1}^{n} |\Im \lambda_k(A)|^2 = N_2^2(A_R) - \sum_{k=1}^{n} |\Re \lambda_k(A)|^2$$

with $A_R := \Re A = (A + A^*)/2$.

3.2   A norm estimate for resolvents

Theorem 3.2. *Let $A \in \mathbf{C}^{n \times n}$. Then*

$$\|R_\lambda(A)\| \le \sum_{k=0}^{n-1} \frac{g^k(A)}{\sqrt{k!} \rho^{k+1}(A, \lambda)} \quad (\lambda \notin \sigma(A)).$$

The proof of this theorem is divided into a series of lemmas which are presented in this section below. *Throughout the book we put $0^0 = 1$.*

Theorem 3.2 is sharp: if A is a normal matrix, then $g(A) = 0$ and we obtain

$$\|R_\lambda(A)\| = \frac{1}{\rho(A, \lambda)}.$$

To prove the theorem we begin with the following

Lemma 3.3. *Let $f(k)$ $(k = 1, ..., n - 1)$ be positive numbers. Then for any positive integer $p < n$ one has*

$$J_p := \sum_{n > k_1 > k_2 > ... > k_p \ge 1} f(k_1) \cdots f(k_p) \le \frac{(\sum_{j=1}^{n-1} f(j))^p}{p!}.$$

Proof. Put

$$f(x) = f(k) \ \ (k - 1 < x \le k; k = 1, ..., n - 1), f(0) = 0.$$

It is obvious that

$$\sum_{k=1}^{j} f(k) \le \int_0^x f(s)ds \ \ (j \le x < j + 1 \le n).$$

So

$$\sum_{j=1}^{n-1} f(j) \sum_{k=1}^{j-1} f(k) \le \int_0^{n-1} f(x) \int_0^x f(s)ds \, dx.$$

Similarly,

$$\sum_{j=1}^{n-1} f(j) \sum_{k=1}^{j-1} f(k) \sum_{i=1}^{k-1} f(i) \le \int_0^{n-1} f(x) \int_0^x f(s) \int_0^s f(y)dy \, ds \, dx.$$

Continuing this process, we get

$$J_p \le \int_0^{n-1} f(x_1) \int_0^{x_1} f(x_2)... \int_0^{x_{p-1}} f(x_p)dx_p \, dx_{p-1} \, ... \, dx_2 \, dx_1.$$

Put

$$z(x) = \int_0^x f(s)ds.$$

Then

$$J_p \le \int_0^{n-1} \int_0^{x_1} ... \int_0^{x_{p-2}} z(x_{p-1})dz(x_{p-1})...dz(x_1) =$$

$$\int_0^{n-1} \int_0^{x_1} ... \int_0^{x_{p-3}} \frac{1}{2} z^2(x_{p-2})dz(x_{p-2})...dz(x_1) = ... = \frac{z^p(n-1)}{p!}.$$

This proves the result. Q. E. D.

Let us estimate powers of nilpotent matrices.

Lemma 3.4. *For any nilpotent operator* $V \in \mathbf{C}^{n \times n}$ *one has*

$$\|V^p\| \le \frac{N_2^p(V)}{\sqrt{p!}} \ \ (p = 1, ..., n - 1).$$

Proof. Since V is nilpotent, due to the Schur theorem we can represent it by an upper-triangular matrix with the zero diagonal:

$$V = (a_{jk})_{j,k=1}^n \text{ with } a_{jk} = 0 \ (j \geq k).$$

Denote

$$\|x\|_m = (\sum_{k=m}^n |x_k|^2)^{1/2} \text{ for } 1 \leq m < n,$$

where x_k $(k = 1, ..., n)$ are coordinates of a vector x. We can write

$$\|Vx\|_m^2 = \sum_{j=m}^{n-1} |\sum_{k=j+1}^n a_{jk}x_k|^2.$$

Hence (by Schwarz's inequality),

$$\|Vx\|_m^2 \leq \sum_{j=m}^{n-1} h_j \|x\|_{j+1}^2, \tag{2.1}$$

where

$$h_j = \sum_{k=j+1}^n |a_{jk}|^2 \ (1 \leq j < n).$$

Furthermore,

$$\|V^2x\|_m^2 = \sum_{j=m}^{n-1} |\sum_{k=j+1}^n a_{jk}(Vx)_k|^2 \leq \sum_{j=m}^{n-1} h_j \|Vx\|_{j+1}^2.$$

Here $(Vx)_k$ are coordinates of Vx. Taking into account (2.1), we obtain

$$\|V^2x\|_m^2 \leq \sum_{j=m}^{n-1} h_j \sum_{k=j+1}^{n-1} h_k \|x\|_{k+1}^2 = \sum_{m \leq j < k \leq n-1} h_j h_k \|x\|_{k+1}^2 \ (1 \leq m \leq n-2).$$

Hence,

$$\|V^2\|^2 \leq \sum_{1 \leq j < k \leq n-1} h_j h_k.$$

Repeating these arguments, we arrive at the inequality

$$\|V^p\|^2 \leq \sum_{1 \leq k_1 < k_2 < ... < k_p \leq n-1} h_{k_1} ... h_{k_p}. \tag{2.2}$$

Therefore due to Lemma 3.3,

$$\|V^p\|^2 \leq \frac{1}{p!}(\sum_{j=1}^{n-1} h_j)^p.$$

But

$$\sum_{j=1}^{n-1} h_j \le N_2^2(V). \tag{2.3}$$

We thus have derived the required result. Q. E. D.

Proof of Theorem 3.2: Due to the triangular representation (1.4),

$$A - \lambda I = D + V - \lambda I = (D - I\lambda)(I + (D - I\lambda)^{-1}V) \quad (\lambda \notin \sigma(A)).$$

Since in the triangular representation of A, D is a diagonal matrix and V is a strictly upper (lower) triangular one, $R_\lambda(D)V$ is a nilpotent operator. So we can write

$$R_\lambda(A) = (I + R_\lambda(D)V)^{-1}R_\lambda(D)$$

$$= \sum_{k=0}^{n-1} (-1)^k (R_\lambda(D)V)^k R_\lambda(D) \quad (\lambda \notin \sigma(A)). \tag{2.4}$$

By virtue of Lemma 3.4,

$$\|(R_\lambda(D)V)^k\| \le \frac{1}{\sqrt{k!}} N_2^k(R_\lambda(D)V) \quad (k = 1, ..., n-1). \tag{2.5}$$

Since D is a normal operator, we have $\|R_\lambda(D)\| = \rho^{-1}(D, \lambda)$. It is clear that

$$N_2(R_\lambda(D)V) \le N_2(V)\|R_\lambda(D)\| = N_2(V)\rho^{-1}(D, \lambda). \tag{2.6}$$

Now (2.4) and (2.5) yield the inequality

$$\|R_\lambda(A)\| \le \sum_{k=0}^{n-1} \frac{N_2^k(V)}{\sqrt{k!}\rho^{k+1}(D, \lambda)}.$$

This relation proves the stated result, since A and D have the same eigenvalues and $N_2(V) = g(A)$, due to Lemma 3.1. Q. E. D.

3.3   Inequality between resolvents and determinants

We begin with the following

Theorem 3.3. *Let $A \in \mathbf{C}^{n \times n}$ be invertible and $1 \le p < \infty$. Then*

$$\|A^{-1} \det A\| \le \frac{N_p^{n-1}(A)}{(n-1)^{(n-1)/p}} \tag{3.1}$$

and, in particular,

$$\|A^{-1} \det A\| \le \|A\|^{n-1}.$$

The proof of this theorem is based on the following

Lemma 3.5. *Let $A \in \mathbf{C}^{n \times n}$ be a strictly positive definite Hermitian matrix. Then*

$$\|A^{-1} \det A\| \leq \left[\frac{\text{trace } A^p}{n-1}\right]^{(n-1)/p} \quad (1 \leq p < \infty)$$

and

$$\|A^{-1} \det A\| \leq \|A\|^{n-1}.$$

Proof. Recall that $\lambda_k(A)$ $(k = 1, ..., n)$ are the eigenvalues taken with their multiplicities. Without loss of generality assume that

$$\lambda_n(A) = \min_{k=1,...,n} \lambda_k(A).$$

Then $\|A^{-1}\| = \lambda_n^{-1}(A)$ and

$$\|A^{-1} \det A\| = \prod_{k=1}^{n-1} \lambda_k(A).$$

Hence,

$$\|A^{-1} \det A\| \leq \|A\|^{n-1}.$$

Moreover, due to the inequality between the arithmetic and geometric mean values we get

$$\|A^{-1} \det A\|^p = \prod_{k=1}^{n-1} \lambda_k^p(A)$$

$$\leq \left[\frac{1}{n-1} \sum_{k=1}^{n-1} \lambda_k^p\right]^{n-1} = [(n-1)^{-1}\text{trace } A^p]^{n-1},$$

as claimed. Q. E. D.

Proof of Theorem 3.3: Let $A \in \mathbf{C}^{n \times n}$ be arbitrary invertible and $A = UB$ the polar representation, where B is positive definite and U is unitary. Then

$$\det A = \det U \det B.$$

Since the absolute values of all the eigenvalues of U equal one we have

$$|\det A| = \det B.$$

Moreover $N_p(B) = N_p(A)$, $\|B\| = \|A\|$ and $\|B^{-1}\| = \|A^{-1}\|$. Now the previous lemma yields

$$\|A^{-1} \det A\| = \|B^{-1}\det B\| \leq \frac{N_p^{n-1}(B)}{(n-1)^{(n-1)/p}} =$$
$$\frac{N_p^{n-1}(A)}{(n-1)^{(n-1)/p}},$$

and

$$\|B^{-1}\det B\| = \|A^{-1} \det A\| \leq \|B\|^{n-1} = \|A\|^{n-1},$$

as claimed. Q. E. D.

Replacing in (3.1) A by $\lambda I - A$ with $p = 2$, we get

$$\|(I\lambda - A)^{-1} \det (\lambda I - A)\| \leq \frac{N_2^{n-1}(\lambda I - A)}{(n-1)^{(n-1)/2}}.$$

Taking into account that trace $I = n$ we can write

$$N_2^2(\lambda I - A) = \text{trace } (\lambda I - A)(\lambda I - A)^* = N_2^2(A) - 2\Re\,(\overline{\lambda}\text{ trace }(A)) + n|\lambda|^2.$$

So we arrive at the following result.

Corollary 3.4. *For any $A \in \mathbf{C}^{n \times n}$ and all regular λ of A one has*

$$\|(I\lambda - A)^{-1} \det (\lambda I - A)\| \leq$$
$$\left[\frac{N_2^2(A) - 2\Re\,(\overline{\lambda}\text{ trace }(A)) + n|\lambda|^2}{n-1}\right]^{(n-1)/2}.$$

This corollary enables us to prove the following result.

Theorem 3.4. *Let $A \in \mathbf{C}^{n \times n}$. Then*

$$\|(I\lambda - A)^{-1})\| \leq \frac{1}{\rho(A,\lambda)}\left[1 + \frac{1}{n-1}\left(1 + \frac{g^2(A)}{\rho^2(A,\lambda)}\right)\right]^{(n-1)/2} \quad (\lambda \notin \sigma(A)),$$

where $\rho(A,\lambda) = \min_k |\lambda_k(A) - \lambda|$.

Proof. We apply equality (2.4), taking into account that $B_\lambda = -R_\lambda(D)V$ is a nilpotent matrix. So Corollary 3.4 yields

$$\|(I + B_\lambda)^{-1}\| \leq \left[1 + \frac{1 + N_2^2(B_\lambda)}{n-1}\right]^{(n-1)/2}.$$

In addition, (2.6) implies

$$\|(I + B_\lambda)^{-1}\| \leq \left[1 + \frac{1}{(n-1)}\left(1 + \frac{N_2^2(V)}{\rho^2(A,\lambda)}\right)\right]^{(n-1)/2}.$$

Take into account that $\|(I\lambda - D)^{-1}\| = \rho^{-1}(D,\lambda)$. Moreover, due to Lemma 3.1, $N_2(V) = g(A)$. Now (2.4) proves the required result. Q. E. D.

3.4 Functions regular on the convex hull of the spectrum

Let $A \in \mathbf{C}^{n \times n}$ and $M \supset \sigma(A)$ be an open simply-connected set whose boundary C consists of a finite number of rectifiable Jordan curves, oriented in the positive sense customary in the theory of complex variables. Suppose that $M \cup C$ is contained in the domain of analyticity of a scalar-valued function f. Recall that

$$f(A) = -\frac{1}{2\pi i} \int_C f(\lambda) R_\lambda(A) d\lambda. \tag{4.1}$$

Theorem 3.5. *Let A be an $n \times n$-matrix and f be a function holomorphic on a neighborhood of the convex hull $co(A)$ of $\sigma(A)$. Then*

$$\|f(A)\| \leq \max_{k=1,\dots,n} |f(\lambda_k(A))| + \sum_{k=1}^{n-1} \sup_{\lambda \in co(A)} |f^{(k)}(\lambda)| \frac{g^k(A)}{(k!)^{3/2}}.$$

If A is normal, then it is required only that f is holomorphic on a neighborhood of $\sigma(A)$. Besides, $\|f(A)\| = \max_{k=1,\dots,n} |f(\lambda_k(A))|$.

This theorem is proved in the next section.

Example 3.2. Let $A \in \mathbf{C}^{n \times n}$. Then

$$\|exp(At)\| \leq e^{\alpha(A)t} \sum_{k=0}^{n-1} \frac{g^k(A)t^k}{(k!)^{3/2}} \quad (t \geq 0),$$

where $\alpha(A) = \max_{k=1,\dots,n} Re\,\lambda_k(A)$.

Example 3.3. Let $A \in \mathbf{C}^{n \times n}$. Then

$$\|A^m\| \leq \sum_{k=0}^{n-1} \frac{m! g^k(A) r_s^{m-k}(A)}{(m-k)!(k!)^{3/2}} \quad (m = 1, 2, \dots),$$

where $r_s(A)$ is the spectral radius.

Note that in the finite dimensional case the following representation for matrix functions is true: consider the $n \times n$-Jordan block

$$J_n(\lambda_0) = \begin{pmatrix} \lambda_0 & 1 & 0 & \dots & 0 \\ 0 & \lambda_0 & 1 & \dots & 0 \\ \cdot & \cdot & \cdot & \dots & \cdot \\ \cdot & \cdot & \cdot & \dots & \vdots \\ \cdot & \cdot & \cdot & \dots & \cdot \\ 0 & 0 & \dots & \lambda_0 & 1 \\ 0 & 0 & \dots & 0 & \lambda_0 \end{pmatrix},$$

then

$$f(J_n(\lambda_0)) = \begin{pmatrix} f(\lambda_0) & \frac{f'(\lambda_0)}{1!} & \cdots & \frac{f^{(n-1)}(\lambda_0)}{(n-1)!} \\ 0 & f(\lambda_0) & \cdots & \\ & \cdot & \cdot & \cdots & \cdot \\ & \cdot & \cdot & \cdots & \cdot \\ & \cdot & \cdot & \cdots & \cdot \\ 0 & \cdots & f(\lambda_0) & \frac{f'(\lambda_0)}{1!} \\ 0 & \cdots & 0 & f(\lambda_0) \end{pmatrix}.$$

Thus, if A has the Jordan block-diagonal form

$$A = diag\ (J_{m_1}(\lambda_1), J_{m_2}(\lambda_2), ..., J_{m_{n_0}}(\lambda_{n_0}))$$

where λ_k, $k = 1, ..., n_0$, are the eigenvalues whose geometric multiplicities are m_k, then

$$f(A) = diag\ (f(J_{m_1}(\lambda_1)), f(J_{m_2}(\lambda_2)), ..., f(J_{m_{n_0}}(\lambda_{n_0}))).$$

Besides, we do not require that f is regular on a neighborhood of $\sigma(A)$: one can take an arbitrary function which has at each λ_k derivatives up to $m_k - 1$-order.

3.5   Proof of Theorem 3.5

Let $\{e_k\}$ be the orthonormal basis of the triangular representation (the Schur basis) of A, and V the nilpotent part of A. Denote by $|V|_e$ the matrix whose entries in $\{e_k\}$ are the absolute values of the entries of V in that basis. That is,

$$|V|_e = \sum_{k=2}^{n} \sum_{j=1}^{k-1} |a_{jk}|(., e_k)e_j,$$

where $a_{jk} = (Ae_k, e_j)$. Put

$$I_{j_1...j_{k+1}} = \frac{(-1)^{k+1}}{2\pi i} \int_C \frac{f(\lambda)d\lambda}{(\lambda_{j_1} - \lambda)...(\lambda_{j_{k+1}} - \lambda)}.$$

Here $\lambda_j = \lambda_j(A)$ are the eigenvalues of A with the multiplicities and enumerated in an arbitrary way. We need the following result.

Lemma 3.6. *Let A be an $n \times n$-matrix and f be a function holomorphic in a Jordan domain containing $\sigma(A)$. Let D be the diagonal part of A. Then*

$$\|f(A) - f(D)\| \leq \sum_{k=1}^{n-1} J_k\| \ |V|_e^k\|,$$

where

$$J_k = \max\{|I_{j_1...j_{k+1}}| : 1 \leq j_1 < ... < j_{k+1} \leq n\}.$$

Proof. By (4.1) and (2.4) we have

$$f(A) - f(D) = -\frac{1}{2\pi i} \int_C f(\lambda)(R_\lambda(A) - R_\lambda(D))d\lambda = \sum_{k=1}^{n-1} B_k, \qquad (5.1)$$

where

$$B_k = (-1)^{k+1} \frac{1}{2\pi i} \int_C f(\lambda)(R_\lambda(D)V)^k R_\lambda(D)d\lambda.$$

Since D is a diagonal matrix with respect to $\{e_k\}$ and its diagonal entries are the eigenvalues of A, we can write

$$R_\lambda(D) = \sum_{j=1}^n \frac{\Delta P_j}{\lambda_j(A) - \lambda},$$

where $\Delta P_k = (., e_k)e_k$. Thus,

$$B_k = \sum_{j_{k+1}=1}^n \sum_{j_k=1}^n \cdots \sum_{j_2=1}^n \sum_{j_1=1}^n \Delta P_{j_1} V \Delta P_{j_2} V \Delta P_{j_3} \ldots \Delta P_{j_k} V \Delta P_{j_{k+1}} I_{j_1 j_2 \ldots j_{k+1}}.$$

In addition, $\Delta P_j V \Delta P_k = 0$ for $j \geq k$. Consequently,

$$B_k =$$

$$\sum_{j_{k+1}=1}^n \sum_{j_k=1}^{j_{k+1}-1} \cdots \sum_{j_2=1}^{j_3-1} \sum_{j_1=1}^{j_2-1} \Delta P_{j_1} V \Delta P_{j_2} V \Delta P_{j_3} \ldots \Delta P_{j_k} V \Delta P_{j_{k+1}} I_{j_1 j_2 \ldots j_{k+1}}.$$

Let $|B_k|_e$ be the operator whose entries in the basis $\{e_k\}$ are the absolute values of the entries of B_k in that basis. Then

$$|B_k|_e \leq J_k \sum_{j_{k+1}=1}^n \sum_{j_k=1}^{j_{k+1}-1} \cdots \sum_{j_2=1}^{j_3-1} \sum_{j_1=1}^{j_2-1} \Delta P_{j_1} |V|_e \Delta P_{j_2} |V|_e \ldots |V|_e \Delta P_{j_{k+1}} =$$

$$J_k P_{n-k} |V|_e P_{n-k+1} |V|_e P_{n-k+2} \ldots P_{n-1} |V|_e,$$

where the inequality is understood in the entry-wise sense with respect to the Schur basis.

But

$$P_{n-k} |V|_e P_{n-k+1} |V|_e P_{n-k+2} \ldots P_{n-1} |V|_e$$

$$= |V|_e P_{n-k+1} |V|_e P_{n-k+2} \ldots P_{n-1} |V|_e$$

$$= |V|_e^2 P_{n-k+2} |V|_e \ldots P_{n-1} |V|_e = \ldots = |V|_e^{k-1} P_{n-1} |V|_e = \ldots = |V|_e^k.$$

Thus

$$|B_k|_e \leq J_k |V|_e^k.$$

Hence,

$$\|B_k\| \leq \||B_k|_e\| \leq J_k \||V|_e^k\|.$$

This inequality and (5.1) imply the required result. Q. E. D.

Due to Lemma 3.4

$$\||V|_e^k\| \leq \frac{N_2^k(|V|_e)}{\sqrt{k!}}.$$

Since $N_2(|V|_e) = N_2(V) = g(A)$, by the previous lemma we arrive at the following result.

Lemma 3.7. *Under the hypothesis of Lemma 3.6 we have*

$$\|f(A) - f(D)\| \leq \sum_{k=1}^{n-1} \frac{g^k(A) J_k}{\sqrt{k!}}.$$

We also need the following lemma.

Lemma 3.8. *Let M_0 be the closed convex hull of points $x_0, x_1, ..., x_n \in \mathbf{C}$ and let a scalar-valued function f be regular on a neighborhood D_1 of M_0. In addition, let $\Gamma \subset D_1$ be a Jordan closed contour surrounding the points $x_0, x_1, ..., x_n$. Then*

$$\left| \frac{1}{2\pi i} \int_\Gamma \frac{f(\lambda) d\lambda}{(\lambda - x_0)...(\lambda - x_n)} \right| \leq \frac{1}{n!} \sup_{\lambda \in M_0} |f^{(n)}(\lambda)|.$$

Proof. First, let all the points be distinct: $x_j \neq x_k$ for $j \neq k$ $(j, k = 0, ..., n)$, and let $D_f(x_0, x_1, ..., x_n)$ be a divided difference of function f at points $x_0, x_1, ..., x_n$. The divided difference admits the representation

$$D_f(x_0, x_1, ..., x_n) = \frac{1}{2\pi i} \int_\Gamma \frac{f(\lambda) d\lambda}{(\lambda - x_0)...(\lambda - x_n)} \tag{5.2}$$

(see [22, Chapter I, formula (54)]). But, on the other hand, the following estimate is well-known:

$$| D_f(x_0, x_1, ..., x_n) | \leq \frac{1}{n!} \sup_{\lambda \in M_0} |f^{(n)}(\lambda)|$$

[22, Chapter I, formula (49)]. Combining that inequality with relation (5.2), we arrive at the required result. If $x_j = x_k$ for some $j \neq k$, then the claimed inequality can be obtained by small perturbations and the previous reasonings. Q. E. D.

Lemma 3.9. *Let $x_0 \leq x_1 \leq ... \leq x_n$ be real points and let a function f be regular on a neighborhood D_1 of the segment $[x_0, x_n]$. In addition, let $\Gamma \subset D_1$ be a Jordan closed contour surrounding $[x_0, x_n]$. Then there is a point $\eta \in [x_0, x_n]$, such that the equality*

$$\frac{1}{2\pi i} \int_\Gamma \frac{f(\lambda) d\lambda}{(\lambda - x_0)...(\lambda - x_n)} = \frac{1}{n!} f^{(n)}(\eta)$$

is true.

Proof. First suppose that all the points are distinct: $x_0 < x_1 < ... < x_n$. Then the divided difference $D_f(x_0, x_1, ..., x_n)$ of f in the points $x_0, x_1, ..., x_n$ admits the representation

$$D_f(x_0, x_1, ..., x_n) = \frac{1}{n!} f^{(n)}(\eta)$$

with some point $\eta \in [x_0, x_n]$ [22, Chapter I, formula (43)], [71, page 5]. Combining this equality with representation (5.2), we arrive at the required result. If $x_j = x_k$ for some $j \neq k$, then the claimed inequality can be obtained by small perturbations and the previous reasonings. Q. E. D.

From Lemma 3.8 it follows

$$J_k \leq \frac{1}{k!} \sup_{\lambda \in co(A)} |f^{(k)}(\lambda)|, k = 1, ..., n-1. \tag{5.3}$$

Now Lemma 3.7 implies

Corollary 3.5. *Under the hypothesis of Theorem 3.5 we have*

$$\|f(A) - f(D)\| \leq \sum_{k=1}^{n-1} \sup_{\lambda \in co(A)} |f^{(k)}(\lambda)| \frac{g^k(A)}{(k!)^{3/2}}.$$

The assertion of Theorem 3.5 directly follows from the previous corollary, since

$$\|f(D)\| = \sup_{\lambda \in \sigma(A)} |f(\lambda)|.$$

3.6   The function $\frac{1}{f}(A)$

Let $f(z)$ be regular on an open simply connected domain $G \subseteq \mathbf{C}$ containing $\sigma(A)$ and

$$f(\lambda) \neq 0 \quad (\lambda \in G). \tag{6.1}$$

Then $\frac{1}{f(\lambda)}$ is regular on $\sigma(A)$. Consider the matrix function

$$\frac{1}{f}(A) := -\frac{1}{2\pi i} \int_C \frac{1}{f(\lambda)} R_\lambda(A) d\lambda, \tag{6.2}$$

where the Jordan contour $C \subset G$ surrounds the spectrum of A. We have

$$f(A)\left(\frac{1}{f}(A)\right) = \left(\frac{1}{f}(A)\right) f(A) = -\frac{1}{2\pi i} \int_C R_\lambda(A) d\lambda = I.$$

So $\frac{1}{f}(A)$ is inverse to $f(A)$. For any invertible $B \in \mathbf{C}^{n \times n}$ from Theorem 3.2 we get

$$\|B^{-1}\| \leq \sum_{k=0}^{n-1} \frac{g^k(B)}{\sqrt{k!}r_{\mathrm{low}}(B)},$$

where $r_{\mathrm{low}}(B) = \rho(B,0) = \min_k |\lambda_k(A)|$ is the lower spectral radius of B. We thus arrive at

Lemma 3.10. *Let f be regular on a neighborhood of $\sigma(A)$ and*

$$f(\lambda_k(A)) \neq 0 \quad (k = 1, ..., n). \tag{6.3}$$

Then

$$\left\|\frac{1}{f}(A)\right\| \leq \sum_{k=0}^{n-1} \frac{g^k(f(A))}{\sqrt{k!}r_{\mathrm{low}}^{k+1}(f(A))},$$

where $r_{\mathrm{low}}(f(A)) = \min_k |f(\lambda_k(A))|$ and

$$g(f(A)) = (N_2^2(f(A)) - \sum_{k=1}^{n} |f(\lambda_k(A))|^2)^{1/2}.$$

To estimate $g(f(A))$ again use the triangular representation, $A = D + V$ $(\sigma(A) = \sigma(D))$, where D the diagonal part and V is the nilpotent part of A (see Section 3.1). It is not hard to show that the matrix

$$V_{f,A} := f(A) - f(D)$$

is the nilpotent part of the matrix $f(A)$ and by Lemma 3.1,

$$N_2(V_{f,A}) = g(f(A)). \tag{6.4}$$

Moreover,

$$g(f(A)) = N_2(V_{f,A}) = N_2(f(A) - f(D)) \le \sqrt{n}\|f(A) - f(D)\|.$$

Making use of Corollary 3.5, we obtain

$$g(f(A)) \le \tilde{g}_{f,A}, \tag{6.5}$$

where

$$\tilde{g}_{f,A} := \sqrt{n} \sum_{j=1}^{n-1} \frac{g^j(A)}{(j!)^{3/2}} \sup_{\lambda \in co(A)} |f^{(j)}(\lambda)|,$$

provided f is regular on $co(A)$. Therefore Lemma 3.10 yields our next result.

Theorem 3.6. *Let f be regular on $co(A)$ and condition (6.3) hold. Then*

$$\left\|\frac{1}{f}(A)\right\| \le \sum_{k=0}^{n-1} \frac{\tilde{g}_{f,A}^k}{\sqrt{k!}r_{\text{low}}^{k+1}(f(A))}.$$

Now we are going to establish a norm estimate for $\frac{1}{f}(A)$ in terms of the determinant of $f(A)$. To this end, supposing that f is holomorphic on a neighborhood of $co(A)$, put

$$\Phi_f(A) := \max_{k=1,\dots,n} |f(\lambda_k(A))| + \sum_{k=1}^{n-1} \sup_{\lambda \in co(A)} |f^{(k)}(\lambda)| \frac{g^k(A)}{(k!)^{3/2}}.$$

Theorem 3.7. *Let f be regular on $co(A)$ and condition (6.3) hold. Then*

$$\left\|\frac{1}{f}(A)\right\| \le \frac{\Phi_f^{n-1}(A)}{|\det f(A)|}.$$

Proof. Due to Theorem 3.3 we have

$$\left\|\frac{1}{f}(A) \det f(A)\right\| = \|(f(A))^{-1} \det f(A)\| \le \|f(A)\|^{n-1}.$$

The required result now is due to Theorem 3.5. Q. E. D.

Let us estimate the norm of A^{-m} for an integer $m \ge 1$. In this case $f(A) = A^m$,

$$g(f(A)) = g(A^m) = (N_2^2(A^m) - \sum_{k=1}^{n} |\lambda_k(A)|^{2m})^{1/2}.$$

Now Lemma 3.10 implies

Corollary 3.6. *Let $r_{\text{low}}(A) > 0$. Then*

$$\|A^{-m}\| \le \sum_{k=0}^{n-1} \frac{g^k(A^m)}{\sqrt{k!}r_{\text{low}}^{m(k+1)}(A)} \quad (m = 1, 2, \dots).$$

This corollary is sharp: if A is normal, then we get the equality

$$\|A^{-m}\| = \frac{1}{r_{\text{low}}^m(A)}.$$

To obtain a norm estimate for A^{-m} in terms of the determinant of A, take into account that

$$|\det A^m| = |\det A|^m,$$

Then Theorem 3.3 implies the next result.

Corollary 3.7. *Let $A \in \mathbf{C}^{n \times n}$ be invertible and $1 \le p < \infty$. Then*

$$\|A^{-m}\| \le \frac{N_p^{n-1}(A^m)}{(n-1)^{(n-1)/p}|\det A|^m}$$

and, in particular,

$$\|A^{-m}\| \le \frac{\|A^m\|^{n-1}}{|\det A|^m}.$$

3.7 Norm estimates for the matrix logarithm

The aim of this section is to illustrate applications of the results obtained above in this chapter to the matrix logarithm defined in Section 1.5.

Lemma 3.11. *Let $A \in \mathbf{C}^{n \times n}$ and*

$$r_0 := \inf\{|z| : z \in co\,(A)\} > 0. \tag{7.1}$$

Then

$$\|\ln(A)\| \le \max_k |\ln(\lambda_k)| + \sum_{k=1}^{n-1} \frac{g^k(A)}{r_0^k k!^{1/2}}. \tag{7.2}$$

Proof. According to (7.1) $\ln(z)$ is regular on $co\,(A)$. Moreover,

$$\left|\frac{d^k}{dz^k}\ln(z)\right| = \frac{(k-1)!}{|z|^k} \le \frac{(k-1)!}{r_0^k} \quad (z \in co(A); \ k = 1, 2, ...).$$

Hence, by Theorem 3.5 we get the required inequality. Q. E. D.

Since the calculations of r_0 is not an easy task, the following result is of some interest.

Lemma 3.12. *Let $A \in \mathbf{C}^{n \times n}$ and the condition*

$$r_s(I - A) < 1 \tag{7.3}$$

hold. Then

$$\|\ln(A)\| \le \sum_{k=0}^{n-1} \sum_{m=1}^{\infty} \frac{m! r_s^{m-k}(A-I)g^k(A)}{m(m-k)!(k!)^{3/2}}.$$

Proof. Let us use the representation

$$\ln(A) = \sum_{k=1}^{\infty} \frac{1}{k} (I - A)^k. \tag{7.4}$$

So

$$\|\ln(A)\| \le \sum_{k=1}^{\infty} \frac{1}{k} \|(I - A)^k\|.$$

By Theorem 3.5, we have

$$\|(I - A)^m\| \le \sum_{k=0}^{n-1} \frac{m! r_s^{m-k} (I - A) g^k (A)}{(m - k)! (k!)^{3/2}} \quad (m = 1, 2, ...).$$

This and (7.4) yield the required inequality. Q. E. D.

Furthermore, suppose that

$$\sigma(A) \cap (-\infty, 0] = \emptyset, \tag{7.5}$$

then we do not require that $\ln(z)$ is regular on $co\,(A)$. As it was shown in Section 1.5, in this case

$$\ln(A) = (A - I) \int_0^{\infty} (tI + A)^{-1} \frac{dt}{1 + t}. \tag{7.6}$$

Hence, due to Theorem 3.2,

$$\|\ln(A)\| \le \|A - I\| \sum_{k=0}^{n-1} \frac{g^k(A)}{\sqrt{k!}} \int_0^{\infty} \frac{dt}{\rho^{k+1}(A, -t)(1 + t)}. \tag{7.7}$$

For instance, let $\sigma(A)$ be purely imaginary and $|\lambda_k(A)| \ge a > 0$ ($k = 1, ..., n$). Then

$$\rho(A, -t) \ge \sqrt{a^2 + t^2}.$$

So in this case,

$$\|\ln(A)\| \le \|A - I\| \int_0^{\infty} \left(\sum_{k=0}^{n-1} \frac{g^k(A)}{\sqrt{k!}(a^2 + t^2)^{(k+1)/2}} \right) \frac{dt}{1 + t}. \tag{7.8}$$

3.8   Spectral representations for resolvents

Again use the triangular representation

$$A = D + V \tag{8.1}$$

with a normal (diagonal) operator D and a nilpotent operator V (see Section 3.1). P_j $(j = 1, ..., n)$ are the invariant projections of A with dim $(P_k - P_{k-1})\mathbf{C}^n = 1$ and $P_0 = 0$.

We can write

$$D = \sum_{k=1}^{n} \lambda_k(A)\Delta P_k,$$

where $\Delta P_k = P_k - P_{k-1}$ $(k = 1, ..., n)$.

Lemma 3.13. *Let Q and V be linear operators in $\mathbf{C}^n$ and let V be a nilpotent operator. Suppose that all the invariant subspaces of V and Q are joint. Then VQ and QV are nilpotent operators.*

Proof. Since all the invariant subspaces of V and Q are the same, these operators have the same basis of the triangular representation. Taking into account that the diagonal entries of V are equal to zero, we easily determine that the diagonal entries of QV and VQ are equal to zero. This proves the required result. Q. E. D.

For linear operators $X_1, X_2, ..., X_m$ acing in $\mathbf{C}^n$ denote

$$\overset{\rightarrow}{\prod_{1 \le k \le m}} X_k := X_1 X_2 ... X_m.$$

That is, the arrow over the symbol of the product means that the indexes of the co-factors increase from left to right.

Theorem 3.8. *For any linear operator A in $\mathbf{C}^n$ we have*

$$I - AR_\lambda(A) = \overset{\rightarrow}{\prod_{1 \le k \le n}} \left(I + \frac{A\Delta P_k}{\lambda - \lambda_k(A)} \right) \quad (\lambda \notin \sigma(A)),$$

where P_k $(k = 1, ..., n)$ is the invariant orthogonal projections of A.

Proof. Denote $E_k = I - P_k$. Since

$$A = (E_k + P_k)A(E_k + P_k) \text{ for any } k = 1, ..., n$$

and $E_1 A P_1 = 0$, we get the relation

$$A = P_1 A E_1 + P_1 A P_1 + E_1 A E_1.$$

Take into account that $\Delta P_1 = P_1$ and $P_1 A P_1 = \lambda_1(A)\Delta P_1$. Then

$$A = \lambda_1(A)\Delta P_1 + \Delta P_1 A E_1 + E_1 A E_1$$

and therefore,

$$A = \lambda_1(A)\Delta P_1 + A E_1. \tag{8.2}$$

Let us check the equality

$$R_\lambda(A) = \Psi(\lambda), \tag{8.3}$$

where

$$\Psi(\lambda) := \frac{\Delta P_1}{\lambda_1(A) - \lambda} - \frac{\Delta P_1}{\lambda_1(A) - \lambda} A E_1 R_\lambda(A) E_1 + E_1 R_\lambda(A) E_1.$$

In fact, multiplying this equality from the left by $A - I\lambda$ and taking into account equality (8.2), we obtain the relation

$$(A - I\lambda)\Psi(\lambda) = \Delta P_1 - \Delta P_1 A E_1 R_\lambda(A) E_1 + (A - I\lambda) E_1 R_\lambda(A) E_1.$$

But $E_1 A E_1 = E_1 A$ and thus $E_1 R_\lambda(A) E_1 = E_1 R_\lambda(A)$. I.e. we can write

$$(A - I\lambda)\Psi(\lambda) = \Delta P_1 + (-\Delta P_1 A + A - I\lambda) E_1 R_\lambda(A) =$$

$$\Delta P_1 + E_1(A - I\lambda) R_\lambda(A) = \Delta P_1 + E_1 = I.$$

Similarly, we multiply (8.3) by $A - I\lambda$ from the right and take into account (8.2). This gives I. Therefore, (8.3) is correct.

Due to (8.3)

$$I - A R_\lambda(A) = (I - (\lambda_1(A) - \lambda)^{-1} A \Delta P_1)(I - A E_1 R_\lambda(A) E_1). \tag{8.4}$$

Now we apply the above arguments to operator $A E_1$. We obtain the following expression which is similar to (8.4):

$$I - A E_1 R_\lambda(A) E_1 =$$

$$(I - (\lambda_2(A) - \lambda)^{-1} A \Delta P_2)(I - A E_2 R_\lambda(A) E_2).$$

For any $k < n$, it similarly follows that

$$I - A E_k R_\lambda(A) E_k =$$

$$(I - \frac{A \Delta P_{k+1}}{\lambda_{k+1}(A) - \lambda})(I - A E_{k+1} R_\lambda(A) E_{k+1}).$$

Substitute this into (8.4) for $k = 1, 2, ..., n-1$. We have

$$I - A R_\lambda(A) = \prod_{1 \leq k \leq n-1}^{\rightarrow} \left(I + \frac{A \Delta P_k}{\lambda - \lambda_k(A)} \right) (I - A E_{n-1} R_\lambda(A) E_{n-1}). \tag{8.5}$$

It is clear that $E_{n-1} = \Delta P_n$. I.e.,

$$I - AE_{n-1}R_\lambda(A)E_{n-1} = I + \frac{A\Delta P_n}{\lambda - \lambda_n(A)}.$$

Now (8.5) implies the required result. Q. E. D.

Let A be a normal matrix. Then

$$A = \sum_{k=1}^{n} \lambda_k(A)\Delta P_k.$$

Hence, $A\Delta P_k = \lambda_k(A)\Delta P_k$. Since $\Delta P_k \Delta P_j = 0$ for $j \neq k$, Theorem 3.8 gives us the equality

$$I - AR_\lambda(A) = I + \sum_{k=1}^{n} \frac{\lambda_k(A)\Delta P_k}{\lambda - \lambda_k(A)}.$$

Consequently,

$$AR_\lambda(A) = A \sum_{k=1}^{n} \frac{\Delta P_k}{\lambda_k(A) - \lambda}.$$

Hence we easily get the well-known spectral representation for the resolvent of a normal matrix.

$$R_\lambda(A) = \sum_{k=1}^{n} \frac{\Delta P_k}{\lambda_k(A) - \lambda}.$$

(Firstly for an invertible normal operator, and then by a small perturbation.) Thus, Theorem 3.8 generalizes the spectral representation for the resolvent of a normal matrix.

Furthermore, the identity

$$I - AR_\lambda(A) = -\lambda R_\lambda(A) \ (\lambda \notin \sigma(A) \cap 0)$$

and Theorem 3.8 imply the equality

$$\lambda R_\lambda(A) = - \overset{\rightarrow}{\prod_{1 \leq k \leq n}} \left(I + \frac{A\Delta P_k}{\lambda - \lambda_k(A)} \right) \ (\lambda \notin \sigma(A) \cap 0), \qquad (8.6)$$

Lemma 3.14. *Let V be a nilpotent operator in $\mathbf{C}^n$ and P_k ($k = 1, ..., n$) be its maximal chain of invariant projections. Then*

$$(I - V)^{-1} = \overset{\rightarrow}{\prod_{2 \leq k \leq n}} (I + V\Delta P_k).$$

Proof. Indeed, all the eigenvalues of V are equal to zero, and $V\Delta P_1 = 0$. Now (8.6) gives us the required relation. Q. E. D.

This lemma allows us to prove the second multiplicative representation of the resolvent of A.

Theorem 3.9. *Let D and V be the diagonal and nilpotent parts, respectively, of operator A acting in $\mathbf{C}^n$. Then*

$$R_\lambda(A) = R_\lambda(D) \prod_{2 \leq k \leq n}^{\rightarrow} \left(I + \frac{V\Delta P_k}{\lambda - \lambda_k(A)} \right) \quad (\lambda \notin \sigma(A)),$$

where P_k, $k = 1, ..., n$, is the maximal chain of invariant projections of A.

Proof. Due to the triangular representation

$$R_\lambda(A) = (A - \lambda I)^{-1} = (D + V - \lambda I)^{-1} = R_\lambda(D)(I + VR_\lambda(D))^{-1}.$$

But $VR_\lambda(D)$ is a nilpotent operator. Take into account that

$$R_\lambda(D)\Delta P_k = (\lambda_k(A) - \lambda)^{-1}\Delta P_k.$$

Now the previous lemma ensures the required relation. Q. E. D.

Making use of the relations $V\Delta P_k = P_{k-1}V\Delta P_k$, $P_{k-1}V^*\Delta P_k = 0$ and

$$V\Delta P_k = 2iP_{k-1}V_I\Delta P_k = 2iP_{k-1}(A_I - \Im\lambda_k(A))\Delta P_k \qquad (8.7)$$

$(A_I = \Im A, V_I = \Im V)$, from the preceding theorem we have

$$R_\lambda(A) = R_\lambda(D) \prod_{2 \leq k \leq n}^{\rightarrow} \left(I + \frac{2iP_{k-1}(A_I - \Im\lambda_k(A))\Delta P_k}{\lambda - \lambda_k(A)} \right) \quad (\lambda \notin \sigma(A)).$$

$$(8.8)$$

3.9 An additional inequality for powers of nilpotent matrices

Let us point an additional norm estimate for the powers of a nilpotent matrix which is sharper than Lemma 3.4 but more cumbersome. To this end, for an integer $n \geq 2$ introduce the numbers

$$\gamma_{n,k} = \sqrt{\frac{C_{n-1}^k}{(n-1)^k}} \quad (k = 1, ..., n-1) \text{ and } \gamma_{n,0} = 1.$$

Here

$$C_n^k = \frac{n!}{(n-k)!k!}$$

are binomial coefficients. Evidently, for all $n > 2$,

$$\gamma_{n,k}^2 = \frac{(n-1)(n-2)\dots(n-k)}{(n-1)^k k!} \leq \frac{1}{k!} \quad (k = 1, 2, \dots, n-1). \tag{9.1}$$

We begin with the following lemma.

Lemma 3.15. *For arbitrary positive numbers* $a_1, \dots, a_n$ *and* $m = 1, \dots, n$, *we have*

$$\sum_{1 \leq k_1 < k_2 < \dots < k_m \leq n} a_{k_1} \dots a_{k_m} \leq \frac{1}{n^m} C_n^m \left[\sum_{k=1}^n a_k\right]^m.$$

Proof. For $m = 1$ the result is obvious. So in this proof it is supposed that $m \geq 2$.

Consider the following function of n positive variables $y_1, \dots, y_n$:

$$R_m(y_1, \dots, y_n) \equiv \sum_{1 \leq k_1 < k_2 < \dots < k_m \leq n} y_{k_1} y_{k_2} \dots y_{k_m}.$$

Let us prove that under the condition

$$\sum_{k=1}^n y_k = n\,b, \tag{9.2}$$

where b is a positive number, function R_m has a unique conditional maximum. To this end denote

$$F_j(y_1, \dots, y_n) \equiv \frac{\partial R_m(y_1, \dots, y_n)}{\partial y_j}.$$

Obviously, $F_j(y_1, \dots, y_n)$ does not depend on y_j, symmetrically depends on other variables, and monotonically increases with respect to each of its variables. A conditional extremum of R_m under (9.2) is a root of the coupled system of the equations

$$F_j(y_1, \dots, y_n) - \lambda \frac{\partial}{\partial y_j} \sum_{k=1}^n y_k = 0 \ (j = 1, \dots, n),$$

where λ is the Lagrange factor. Therefore,

$$F_j(y_1, \dots, y_n) = \lambda \ (j = 1, \dots, n).$$

Since $F_j(y_1, \dots, y_n)$ does not depend on y_j, and $F_k(y_1, \dots, y_n)$ does not depend on y_k, equality

$$F_j(y_1, \dots, y_n) = F_k(y_1, \dots, y_n) = \lambda$$

for all $k \neq j$ is possible if and only if $y_j = y_k$. Thus R_m has under (9.2) a unique extremum when

$$y_1 = y_2 = ... = y_n = b. \tag{9.3}$$

But

$$R_m(b, ..., b) = b^m \sum_{1 \leq k_1 < k_2 < ... < k_m \leq n} 1 = b^m C_n^m.$$

Let us check that (9.3) gives us the maximum. Letting

$$y_1 \to nb \text{ and } y_k \to 0 \ (k = 2, ..., n),$$

we get

$$R_m(y_1, ..., y_n) \to 0.$$

Since the extremum (9.3) is unique, it is the maximum. Thus, under (9.2)

$$R_m(y_1, ..., y_n) \leq b^m C_n^m \ (y_k \geq 0, \ k = 1, ..., n).$$

Take $y_j = a_j$ and

$$b = \frac{a_1 + ... + a_n}{n}.$$

Then

$$R_m(a_1, ..., a_n) \leq C_n^m n^{-m} [\sum_{k=1}^{n} a_k]^m.$$

We thus get the required result. Q. E. D.

The latter lemma gives us an additional estimate for the powers of nilpotent matrices.

Lemma 3.16. *For any nilpotent operator $V \in \mathbf{C}^{n \times n}$ one has*

$$\|V^k\| \leq \gamma_{n,k} N_2^k(V) \ (k = 1, ..., n-1).$$

Proof. Due to (2.2),

$$\|V^p\|^2 \leq \sum_{1 \leq k_1 < k_2 < ... < k_p \leq n-1} h_{k_1} ... h_{k_p}.$$

Now Lemma 3.15 yields the inequality

$$\|V^k\|^2 \leq \gamma_{n,k}^2 (\sum_{j=1}^{n-1} h_j)^k.$$

But according to (2.3),

$$\sum_{j=1}^{n-1} h_j \leq N_2^2(V).$$

We thus have derived the required result. Q. E. D.

Lemma 3.16 enables us to obtain the following result which is slightly sharper but more cumbersome than Theorem 3.2.

Theorem 3.10. *Let* $A \in \mathbf{C}^{n \times n}$. *Then*

$$\|R_\lambda(A)\| \leq \sum_{k=0}^{n-1} \frac{g^k(A)\gamma_{n,k}}{\rho^{k+1}(A,\lambda)} \quad (\lambda \notin \sigma(A)).$$

Proof. Due to (2.4)

$$\|R_\lambda(A)\| \leq \sum_{k=0}^{n-1} \|(R_\lambda(D)V)^k\| \|R_\lambda(D)\|.$$

Since $R_\lambda(D)V$ is nilpotent, the preceding lemma yields

$$\|R_\lambda(A)\| \leq \sum_{k=0}^{n-1} \gamma_{n,k} N_2^k(R_\lambda(D)V) \|R_\lambda(D)\| \leq \sum_{k=0}^{n-1} \gamma_{n,k} \|R_\lambda(D)\|^{k+1} N_2^k(V).$$

This relation proves the stated result, since $\|R_\lambda(D)\| \leq \rho^{-1}(D,\lambda)$, A and D have the same eigenvalues and $N_2(V) = g(A)$ due to Lemma 3.1.
Q. E. D.

In addition, Lemma 3.16 allows us to derive a norm estimate for matrix functions, which is slightly sharper but more cumbersome than Theorem 3.5. Namely, the following theorem is valid.

Theorem 3.11. *Let* A *be an* $n \times n$-*matrix and* f *be a function holomorphic on a neighborhood of the convex hull* $co(A)$ *of* $\sigma(A)$. *Then*

$$\|f(A)\| \leq \sup_{\lambda \in \sigma(A)} |f(\lambda)| + \sum_{k=1}^{n-1} \sup_{\lambda \in co(A)} |f^{(k)}(\lambda)| \frac{\gamma_{n,k} g^k(A)}{k!}.$$

Proof. Making use of Lemmas 3.6 and 3.16 we have

$$\|f(A) - f(D)\| \leq \sum_{k=1}^{n-1} J_k \| |V|_e^k \| \leq \sum_{k=1}^{n-1} J_k \gamma_{n,k} N_2^k(|V|_e).$$

But due to Lemma 3.1 $N_2(|V|_e) = N_2(V) = g(A)$. Now inequality (5.3) proves the theorem. Q. E. D.

3.10 Perturbations of matrices

In this section we estimate the spectral variation of two matrices and rotation of simple eigenvectors.

Theorem 3.2 and Lemma 1.10 imply

Corollary 3.8. *Let A and $\tilde{A}$ be $n \times n$-matrices and $q = \|\tilde{A} - A\|$. Then* $\mathrm{sv}_A(\tilde{A}) \le z_n(A, q)$, *where $z_n(A, q)$ is the unique positive root of the equation*

$$q \sum_{j=0}^{n-1} \frac{g^j(A)}{\sqrt{j!}z^{j+1}} = 1. \tag{10.1}$$

Since $g(A) \le \sqrt{2}N_2(A_I)$ (due to Theorem 3.1), one can replace $g(A)$ by $\sqrt{2}N_2(A_I)$.

If A is normal, then $g(A) = 0$, $p(A, z) = z^{n-1}$, $z(A, q) = q$ and therefore, *Corollary 3.8 gives us the well known inequality* $\mathrm{sv}_A(\tilde{A}) \le q$, cf. [6, 76].

Equation (10.1) can be written as

$$z^n = q \sum_{j=0}^{n-1} \frac{g^j(A)}{\sqrt{j!}} z^{n-j-1}. \tag{10.2}$$

To estimate $z_n(A, q)$ one can apply the well known bounds for the roots of polynomials. For instance, consider the algebraic equation

$$z^n = p(z) \quad (n > 1), \text{ where } p(z) = \sum_{j=0}^{n-1} c_j z^{n-j-1} \tag{10.3}$$

with non-negative coefficients c_j $(j = 0, ..., n - 1)$.

Lemma 3.17. *The unique positive root $\hat{z}_0$ of (10.3) satisfies the inequality*

$$\hat{z}_0 \le \begin{cases} p(1) & \text{if } p(1) > 1, \\ p^{1/n}(1) & \text{if } p(1) \le 1. \end{cases}$$

Proof. Since all the coefficients of $p(z)$ are non-negative, it does not decrease as $z > 0$ increases. If $p(1) \le 1$, then $\hat{z}_0 \le 1$ and $p(\hat{z}_0) \le p(1)$. Hence $\hat{z}_0^n \le p(1)$. If $p(1) \ge 1$, then $\hat{z}_0 \ge 1$, $\hat{z}_0^n = p(\hat{z}_0) \le \hat{z}_0^{n-1}p(1)$ and $\hat{z}_0 \le p(1)$, as claimed. Q. E. D.

Substitute $z = g(A)x$ into (10.2), assuming that $g(A) \ne 0$. Then we obtain the equation

$$x^n = \frac{q}{g(A)} \sum_{j=0}^{n-1} \frac{x^{n-j-1}}{\sqrt{j!}}. \tag{10.4}$$

Putting

$$\hat{p}_n = \sum_{j=0}^{n-1} \frac{1}{\sqrt{j!}}$$

and applying Lemma 3.17, for the unique positive root x_0 of (10.4), we obtain

$$x_0 \leq \begin{cases} \frac{q\hat{p}_n}{g(A)} & \text{if } q\hat{p}_n > g(A), \\ \left(\frac{q\hat{p}_n}{g(A)}\right)^{1/n} & \text{if } q\hat{p}_n \leq g(A). \end{cases}$$

But $z_n(A, q) = g(A)x_0$; consequently, according to Corollary 3.8 we get

$$\text{sv}_A(\tilde{A}) \leq \begin{cases} q & \text{if } q\hat{p}_n > g(A), \\ (q\hat{p}_n)^{1/n}g^{1-1/n}(A) & \text{if } q\hat{p}_n \leq g(A). \end{cases} \tag{10.5}$$

Furthermore, put

$$\hat{g}(\tilde{A}, A) = \max\{g(A), g(\tilde{A})\}.$$

Then Corollary 3.8 implies

Corollary 3.9. *One has* $hd(A, \tilde{A}) \leq \hat{z}(A, \tilde{A})$, *where* $\hat{z}(A, \tilde{A})$ *is the unique positive root of the equation*

$$z^n = q \sum_{j=0}^{n-1} \frac{\hat{g}^j(\tilde{A}, A)}{\sqrt{j!}z^{j+1}} z^{n-1-k}.$$

To estimate $\hat{z}(A, \tilde{A})$ one can apply Lemma 3.17.

Now consider the rotation of simple eigenvectors of finite matrices under perturbations. For a $b \in \mathbf{C}$ and an $r > 0$, let $\Omega(b, r) := \{z \in \mathbf{C} : |z-b| < r\}$, and A has a simple eigenvalue μ. So

$$\chi(\mu) := \frac{1}{2} \inf_{s \in \sigma(A), s \neq \mu} |s - \mu| > 0.$$

Put

$$\zeta_n(A, a) = \sum_{j=0}^{n-1} \frac{g^j(A)}{\sqrt{j!}a^{j+1}} \quad (a > 0),$$

assume that

$$q\zeta_n(A, \chi(\mu)) < 1 \tag{10.6}$$

and denote

$$\eta_n(q, A) = \frac{q\chi(\mu)\zeta_n^2(A, \chi(\mu))}{1 - q\zeta_n(A, \chi(\mu))}.$$

Theorem 3.2 and Lemma 1.14 imply

Corollary 3.10. *Let $A, \tilde{A} \in \mathbf{C}^{n \times n}$, and A have a simple eigenvalue μ. If the conditions (10.6) and $\eta_n(q, A) < 1$ hold, then for the normed eigenvector e of A corresponding to μ, there exists a normed eigenvector $\tilde{e}$ of $\tilde{A}$, corresponding to the simple eigenvalue $\lambda(\tilde{A}) \in \Omega(\mu, \chi(A))$, such that*

$$\|e - \tilde{e}\| \leq \frac{2\eta_n(q, A)}{1 - \eta_n(q, A)}.$$

3.11 Comments to Chapter 3

The material in this chapter is based on Chapter 2 of the book [28]. The first time Theorems 3.2 and 3.5 appear in [24] and [25], respectively. Applications of the results presented in this chapter to ordinary differential, difference, differential-delay and neutral type functional differential equations are discussed in the books [29, 30, 35] and [36], respectively. The relevant results can be found in [27].

Chapter 4

Solution Estimates for Polynomial Matrix Equations

In this chapter we consider the Sylvester, quasi-Sylvester, polynomial and generalized polynomial equations in Euclidean spaces. Solution estimates in terms of the spectral (operator) norm are established. We also derive norm estimates for a class of functions of two matrix arguments.

Throughout this chapter $A \in \mathbf{C}^{n \times n}$, $B \in \mathbf{C}^{\tilde{n} \times \tilde{n}}$ and $C \in \mathbf{C}^{n \times \tilde{n}}$ ($n, \tilde{n} < \infty$).

4.1 The quasi-Sylvester matrix equation

Consider the equation

$$Y - AYB = C. \tag{1.1}$$

Due to Lemma 2.1, under the condition

$$r_s(A)r_s(B) < 1, \tag{1.2}$$

equation (1.1) has a unique solution Y which can be represented as

$$Y = \sum_{k=0}^{\infty} A^k C B^k. \tag{1.3}$$

Recall that $1/(t-k)! = 0$ if $t < k$.

Theorem 4.1. *Let condition (1.2) hold. Then the unique solution Y of (1.1) is subject to the inequality*

$$\|Y\| \leq \|C\| \sum_{j=0}^{\tilde{n}-1} \sum_{k=0}^{n-1} \frac{g^k(A)g^j(B)}{(k!j!)^{3/2}} \sum_{t=0}^{\infty} \frac{(t!)^2 r_s^{t-k}(A) r_s^{t-j}(B)}{(t-k)!(t-j)!}. \tag{1.4}$$

73

Proof. By Example 3.3 we can write

$$\left\| \sum_{t=0}^{\infty} A^t C B^t \right\| \le \|C\| \sum_{t=0}^{\infty} \|A^t\| \|B^t\|$$

$$\le \|C\| \sum_{t=0}^{\infty} \sum_{k=0}^{n-1} \frac{t! g^k(A) r_s^{t-k}(A)}{(t-k)!(k!)^{3/2}} \sum_{j=0}^{\tilde{n}-1} \frac{t! g^j(B) r_s^{t-j}(B)}{(t-j)!(j!)^{3/2}}$$

$$= \|C\| \sum_{j=0}^{\tilde{n}-1} \sum_{k=0}^{n-1} \frac{g^k(A) g^j(B)}{(k! j!)^{3/2}} \sum_{t=0}^{\infty} \frac{(t!)^2 r_s^{t-k}(A) r_s^{t-j}(B)}{(t-k)!(t-j)!}.$$

The series

$$\sum_{t=0}^{\infty} \frac{(t!)^2 r_s^t(A) r_s^t(B)}{(t-k)!(t-j)!}$$

converges, since

$$\lim_{t \to \infty} \frac{[(t+1)!]^2 (t-k)!(t-j)!}{(t!)^2 (t+1-k)!(t+1-j)!} = \lim_{t \to \infty} \frac{(t+1)^2}{(t+1-k)(t+1-j)} = 1$$

and (1.2) holds. Now the required inequality follows from (1.3). Q. E. D.

If A is normal, then $g(A) = 0$ and thus (1.4) yields the inequality

$$\|Y\| \le \|C\| \sum_{j=0}^{\tilde{n}-1} \frac{g^j(B)}{(j!)^{3/2}} \sum_{t=0}^{\infty} \frac{t! r_s^t(A) r_s^{t-j}(B)}{(t-j)!}.$$

Now we are going to estimate a solution to (1.1) via an integral. To this end assume that there is a $b \in \mathbf{C}, b \neq 0$, such that

$$|b| r_s(A) < 1 \text{ and } r_s(B) < |b|. \tag{1.5}$$

Then due to Lemma 2.2 the unique solution Y to (1.1) is representable by the integral

$$Y = \frac{b}{2\pi} \int_0^{2\pi} (I e^{-i\omega} - bA)^{-1} C (b e^{i\omega} I - B)^{-1} d\omega. \tag{1.6}$$

Let

$$\zeta_n(A, \delta) = \sum_{j=0}^{n-1} \frac{g^j(A)}{\sqrt{j!} \delta^{j+1}} \quad (\delta > 0).$$

Similarly $\zeta_{\tilde{n}}(B, \delta)$ is defined. Due to Theorem 3.2,

$$\|(A - \lambda I)^{-1}\| \le \zeta_n(A, \rho(A, \lambda)).$$

It is clear that

$$\rho(bA, e^{-i\omega}) \geq 1 - r_s(bA) = 1 - |b|r_s(A)$$

and

$$\rho(B, be^{i\omega}) \geq |b| - r_s(B).$$

From (1.6) we obtain

Corollary 4.1. *Let condition (1.5) hold. Then the unique solution of (1.1) satisfies the inequality*

$$\|Y\| \leq |b|\|C\|\zeta_n(A, 1 - |b|r_s(A))\zeta_{\tilde{n}}(B, |b| - r_s(B)).$$

4.2 The Sylvester equation

Consider the equation

$$AX - XB = C, \tag{2.1}$$

assuming that

$$\sigma(A) \cap \sigma(B) = \emptyset. \tag{2.2}$$

Due to Theorem 2.3,

$$X = \frac{1}{2\pi i} \int_\Gamma (A - \lambda I)^{-1} C (B - \lambda I)^{-1} d\lambda,$$

where Γ is a union of closed contours in the plane, with total winding numbers around $\sigma(A)$ and 0 around $\sigma(B)$. Hence,

$$\|X\| \leq \frac{l_\Gamma \|C\|}{2\pi} \sup_{\lambda \in \Gamma} (\|(A - \lambda I)^{-1}\|\|(B - \lambda I)^{-1}\|)$$

where $l_\Gamma = \int_\Gamma |d\lambda|$ is the length of Γ. Now Theorem 3.2 implies

Corollary 4.2. *Let condition (2.2) hold. Then the solution X of (2.1) satisfies the inequality*

$$\|X\| \leq \frac{l_\Gamma \|C\|}{2\pi} \zeta_n(A, \delta_A)\zeta_{\tilde{n}}(B, \delta_B),$$

where

$$\delta_A = \text{dist } (\Gamma, \sigma(A)) = \inf_{\lambda \in \Gamma, s \in \sigma(A)} |\lambda - s|, \delta_B = \text{dist } (\Gamma, \sigma(B)).$$

Now we are going to consider the Sylvester equation under the condition

$$\beta(B) > \alpha(A). \tag{2.3}$$

To this end put

$$\gamma(a, A, B) := \sum_{j=0}^{\tilde{n}-1} \sum_{k=0}^{n-1} \frac{(k+j)! g^k(A) g^j(B)}{a^{k+j+1}(k! j!)^{1/2}}$$

for a constant $a > 0$.

Lemma 4.1. *Let condition (2.3) hold. Then*

$$\int_0^\infty \|e^{At}\| \|e^{-Bt}\| dt \le \gamma(\eta, A, B),$$

where $\eta := \beta(B) - \alpha(A)$.

Proof. Due to Example 3.2 we have

$$\|e^{-Bs}\| \le e^{-\beta(B)s} \sum_{k=0}^{n-1} \frac{s^k g^k(B)}{(k!)^{3/2}}$$

and

$$\|e^{sA}\| \le e^{\alpha(A)s} \sum_{j=0}^{\tilde{n}-1} \frac{s^j g^j(A)}{(j!)^{3/2}}$$

for all $s \ge 0$. So

$$\|e^{As}\| \|e^{-Bs}\| \le e^{-\eta s} \sum_{j=0}^{\tilde{n}-1} \sum_{k=0}^{n-1} \frac{s^{k+j} g^k(A) g^j(B)}{(k! j!)^{3/2}}.$$

But

$$\int_0^\infty s^{k+j} e^{-\eta s} ds = \frac{(k+j)!}{\eta^{k+j+1}}.$$

This proves the lemma. Q. E. D.

Theorem 4.2. *Let condition (2.3) hold. Then the unique solution to equation (2.1) is subject to the inequality*

$$\|X\| \le \|C\| \gamma(\eta, A, B). \tag{2.4}$$

This result is due to Theorem 2.4 and the previous lemma.

If A is normal, then $g(A) = 0$ and therefore,

$$\gamma(\eta, A, B) = \sum_{j=0}^{n-1} \frac{g^j(B)}{\eta^{j+1}(j!)^{1/2}} = \zeta_n(B, \eta).$$

If both A and B are normal, then $\gamma(\eta, A, B) \le \frac{1}{\eta}$. It is not hard to check that (2.4) is attained: $\gamma(\eta, A, B) = \frac{1}{\eta}$, if A, B and C are commuting normal matrices.

4.3 Polynomial matrix equations

In this section we consider the equations

$$\sum_{k=0}^{m} c_{m-k} A^k X B^k = C,\qquad (3.1)$$

and

$$\sum_{k=0}^{m} c_{m-k} A^k X B^{m-k} = C,\qquad (3.2)$$

where $c_k \in \mathbf{C}$ $(k = 1, ..., m)$, $c_0 = 1$, $A \in \mathbf{C}^{n \times n}, B \in \mathbf{C}^{\tilde{n} \times \tilde{n}}, C \in \mathbf{C}^{n \times \tilde{n}}$.

Let x_k $(k = 1, ..., m)$ be the roots taken with the multiplicities of the polynomial

$$p(x) = \sum_{k=0}^{m} c_k x^{m-k}.$$

4.3.1 *Equation (3.1)*

Let

$$r_s(A) r_s(B) < \min_k |x_k|.\qquad (3.3)$$

Then due to Lemma 2.4 equation (3.1) has a unique solution defined by

$$X = \prod_{k=1}^{m} (A_l B_r - x_k I)^{-1} C.\qquad (3.4)$$

Furthermore, for an $a \in \mathbf{C}$, satisfying

$$|a| > r_s(A) r_s(B),\qquad (3.5)$$

due to Lemma 2.2 we have

$$(A_l B_r - aI)^{-1} C = -\sum_{k=0}^{\infty} \frac{1}{a^{k+1}} A_l^k B_r^k C = -\sum_{k=0}^{\infty} \frac{1}{a^{k+1}} A^k C B^k.\qquad (3.6)$$

Hence and from Example 3.3 it directly follows

$$\|(A_l B_r - aI)^{-1}\| \le J(A, B, a),$$

where

$$J(A, B, a) = \sum_{j=0}^{\tilde{n}-1} \sum_{k=0}^{n-1} \frac{g^k(A) g^j(B)}{(k! j!)^{3/2}} \sum_{w=0}^{\infty} \frac{1}{|a|^{w+1}} \frac{(w!)^2 r_s^{w-k}(A) r_s^{w-j}(B)}{(w-k)!(w-j)!}.$$

Now (3.4) implies

Corollary 4.3. *Let condition (3.3) hold. Then a unique solution X to equation (3.1) satisfies the inequality*

$$\|X\| \le \|C\| J(A, B, x_1) J(A, B, x_2) \cdots J(A, B, x_m).$$

If A is normal, then $g(A) = 0$ and

$$J(A, B, a) = \sum_{j=0}^{\tilde{n}-1} \frac{g^j(B)}{(j!)^{3/2}} \sum_{w=0}^{\infty} \frac{1}{|a|^{w+1}} \frac{w! r_s^w(A) r_s^{w-j}(B)}{(w-j)!}.$$

4.3.2   Equation (3.2)

Consider equation (3.2) supposing that

$$\min_k \beta(x_k B) > \alpha(A). \tag{3.7}$$

According to Lemma 4.2, we have

$$\|(A_l - x_k B_r)^{-1}\| \leq \gamma(\eta_k, A, B),$$

where $\eta_k = \beta(x_k B) - \alpha(A)$. Making use of Theorem 2.8, we arrive at our next result.

Corollary 4.4. *Let condition (3.7) hold. Then a unique solution X to equation (3.2) in $\mathbf{C}^n$ satisfies the inequality*

$$\|X\| \leq \|C\| \gamma(\eta_1, A, B) \gamma(\eta_2, A, B) \cdots \gamma(\eta_m, A, B).$$

4.4   Functions of two non-commuting matrices

By $co(A)$ we again denote the closed convex hull of $\sigma(A)$. Let Ω_A and Ω_B be neighborhoods of $co(A)$ and $co(B)$, respectively. Let $f(z, w)$ be regular on $\Omega_A \times \Omega_B$ and

$$F(f, A, C, B) = -\frac{1}{4\pi^2} \int_{\Gamma_B} \int_{\Gamma_A} f(z, w) R_z(A) C R_w(B) dw \, dz, \tag{4.1}$$

where $\Gamma_A \subset \Omega_A, \Gamma_B \subset \Omega_B$ are closed Jordan contours surrounding $\sigma(A)$ and $\sigma(B)$, respectively.

Define numbers $\psi_{jk} = \psi_{jk}(f, A, B)$ by

$$\psi_{00} = \sup_{z \in \sigma(A), w \in \sigma(B)} |f(z, w)|; \; \psi_{jk} = \frac{1}{(j! k!)^{3/2}} \sup_{z \in co(A), w \in co(B)} \left| \frac{\partial^{j+k} f(z, w)}{\partial z^j \partial w^k} \right|;$$

$$\psi_{0k} := \frac{1}{(k!)^{3/2}} \sup_{z \in \sigma(A), w \in co(B)} \left| \frac{\partial^k f(z, w)}{\partial w^k} \right|,$$

and

$$\psi_{j0} := \frac{1}{(j!)^{3/2}} \sup_{z \in co(A), w \in \sigma(B)} \left| \frac{\partial^j f(z, w)}{\partial z^j} \right|.$$

$(j = 1, ..., n - 1; k = 1, ..., \tilde{n} - 1).$

Theorem 4.3. *Let* $A \in \mathbf{C}^{n \times n}$, $B \in \mathbf{C}^{\tilde{n} \times \tilde{n}}$, $C \in \mathbf{C}^{n \times \tilde{n}}$ *and* $f(z, w)$ *be regular on a neighborhood of* $co(A) \times co(B)$, *then*

$$\|F(f, A, C, B)\| \leq N_2(C) \sum_{j=0}^{n-1} \sum_{k=0}^{\tilde{n}-1} \psi_{jk} g^j(A) g^k(B).$$

If A *is normal and* $f(z, w)$ *is regular on a neighborhood of* $\sigma(A) \times co(B)$, *then*

$$\|F(f, A, C, B)\| \leq N_2(C) \sum_{j=0}^{\tilde{n}-1} \psi_{0j} g^j(B).$$

If both A *and* B *are normal and* $f(z, w)$ *is regular on a neighborhood of* $\sigma(A) \times \sigma(B)$, *then*

$$\|F(f, A, C, B)\| \leq N_2(C) \sup_{z \in \sigma(A), w \in \sigma(B)} |f(z, w)|.$$

This theorem is proved in the next section.

4.5 Proof of Theorem 4.3

Put

$$f^{(j,k)}(z, w) = \left| \frac{\partial^{j+k} f(z, w)}{\partial z^j \partial w^k} \right| \quad (j, k \geq 0).$$

We begin with the following lemma.

Lemma 4.2. *Let* Ω *and* $\tilde{\Omega}$ *be the closed convex hulls of the complex points* $x_0, x_1, ..., x_n$ *and* $y_0, y_1, ..., y_m$, *respectively. Let* M *and* $\tilde{M}$ *be neighborhoods of* Ω *and* $\tilde{\Omega}$, *respectively, and let a scalar-valued function* $f(z, w)$ *be regular on* $M \times \tilde{M}$. *Additionally, let* L *and* $\tilde{L}$ *be the boundaries of* M *and* $\tilde{M}$, *respectively. Then with the notation*

$$\zeta(x_0, ..., x_n; y_0, ...y_m) :=$$

$$-\frac{1}{4\pi^2} \int_L \int_{\tilde{L}} \frac{f(z, w) dz \, dw}{(z - x_0) \cdots (z - x_n)(w - y_0) \cdots (w - y_m)},$$

we have

$$|\zeta(x_0, ..., x_n; y_0, ...y_m)| \leq \frac{1}{n! m!} \sup_{z \in \Omega, w \in \tilde{\Omega}} |f^{(n,m)}(z, w)|.$$

Proof. First, let $x_j \neq x_k, y_j \neq y_k$ for $j \neq k$. Let a function h of one variable be regular on D and $[x_0, x_1, ..., x_n]h$ be a divided difference of function h at points $x_0, x_1, ..., x_n$. Then

$$[x_0, x_1, ..., x_n]h = \frac{1}{2\pi i} \int_L \frac{h(\lambda)d\lambda}{(\lambda - x_0) \cdots (\lambda - x_n)} \qquad (5.1)$$

(see [22, Chapter 1, formula (54)]). Thus

$$\zeta(x_0, ...x_n; \ y_0, ..., y_m) = \frac{1}{2\pi i} \int_{\tilde{L}} \frac{[x_0, ..., x_n]f(., w)dw}{(w - y_0) \cdots (w - y_m)}.$$

Now apply (5.1) to $[x_0, ..., x_n]f(., w)$. Then

$$\zeta(x_0, ...x_n; \ y_0, ..., y_m) = [x_0, ...x_n] \, [y_0, ...y_m]f(.,.)$$

$$\equiv [x_0, ...x_n] \, ([y_0, ...y_m]f(.,.)).$$

The following estimate is well-known [71, p. 6]:

$$|[x_0, ...x_n] \, [y_0, ...y_m]f(.,.)| \leq \sup_{z \in \Omega, w \in \tilde{\Omega}} |f^{(n,m)}(z, w)|.$$

It proves the required result if all the points are distinct. If some points coincide, then the claimed inequality can be obtained by small perturbations and the previous arguments. Q. E. D.

Let $e = \{e_k\}_{k=1}^n$ and $\tilde{e} = \{\tilde{e}_k\}_{k=1}^{\tilde{n}}$ be the orthogonal normal bases of the triangular representation (Schur's bases) to A and B respectively. So

$$Ae_k = \sum_{j=1}^k a_{jk}e_j, \ B\tilde{e}_k = \sum_{j=1}^k \tilde{a}_{jk}\tilde{e}_j \ \ (\tilde{a}_{jk}, a_{jk} \in \mathbf{C}).$$

We can write

$$A = D_A + V_A, B = D_B + V_B, \qquad (5.2)$$

where D_A, D_B are the diagonal parts, V_A and V_B are the nilpotent parts of A and B, respectively. Namely,

$$D_A e_k = \lambda_k(A)e_k \ (\lambda_k(A) \in \sigma(A); k = 1, ..., n)$$

and

$$V_A e_k = \sum_{j=1}^{k-1} a_{jk}e_j \ (k = 2, ..., n); V_A e_1 = 0.$$

Similarly, D_B and V_B are defined. Furthermore, let $|V_A|_e$ be the operator whose entries in $e = \{e_k\}$ are the absolute values of the entries of the matrix V_A. That is, $(|V_A|_e e_j, e_k) = |(V_A e_j, e_k)|$ and

$$|V_A|_e = \sum_{k=1}^{n} \sum_{j=1}^{k-1} |a_{jk}|(., e_k)e_j.$$

Similarly $|V_B|_{\tilde{e}}$ is defined with respect to $\tilde{e} = \{\tilde{e}_k\}$. In addition, $|C|$ is defined by

$$|C|\tilde{e}_j = \sum_{k=1}^{n} |(C\tilde{e}_j, e_k)|e_k.$$

We need the following result

Lemma 4.3. *Let* $f(z, w)$ *be regular on a neighborhood of* $co(A) \times co(B)$. *Then*

$$\|F(f, A, C, B)\| \leq \||C|\| \sum_{j=0}^{n-1} \sum_{k=0}^{\tilde{n}-1} \sqrt{k!j!}\psi_{jk}\||V_A|_e^j\|\ \||V_B|_{\tilde{e}}^k\|.$$

Proof. The triangular representation (5.2) implies the equality

$$(A - I\lambda)^{-1} = (D_A + V_A - \lambda I)^{-1} = (I + R_\lambda(D_A)V_A)^{-1} R_\lambda(D_A)$$

for all regular λ. V_A and $R_\lambda(D_A)$ have the joint invariant subspaces and V_A is nilpotent. So in the triangular representation $R_\lambda(D_A)V_A$ has the zero diagonal. Thus it is nilpotent and consequently, $(R_\lambda(D_A)V_A)^n = 0$. Therefore,

$$(I + R_\lambda(D_A)V_A)^{-1} = \sum_{j=0}^{n-1} (-1)^j (R_\lambda(D_A)V_A)^j$$

and from (4.1) it follows

$$F(f, A, C, B) = \sum_{j=0}^{n-1} \sum_{k=0}^{\tilde{n}-1} M_{jk}, \tag{5.3}$$

where

$$M_{jk} = \frac{(-1)^{k+j+1}}{4\pi^2}$$

$$\times \int_{\Gamma_B} \int_{\Gamma_A} f(z, w)(R_z(D_A)V_A)^j R_z(D_A)C(R_w(D_B)V_B)^k R_w(D_B)dz\ dw$$

for $j < n, k < \tilde{n}$. Since D_A is a diagonal matrix with respect to the Schur basis $\{e_k\}$ and its diagonal entries are the eigenvalues of A, we obtain

$$R_z(D_A) = \sum_{j=1}^{n} \frac{Q_j}{\lambda_j(A) - z},$$

where $Q_k = (., e_k)e_k$. Similarly,

$$R_z(D_B) = \sum_{j=1}^{\tilde{n}} \frac{\tilde{Q}_j}{\lambda_j(B) - z} \text{ where } \tilde{Q}_k = (., \tilde{e}_k)\tilde{e}_k.$$

Taking into account that $Q_s V_A Q_m = 0$, $\tilde{Q}_s V_B \tilde{Q}_m = 0$ $(s \geq m)$, we get

$$M_{jk} = \sum_{1 \leq s_1 < s_2 < ... < s_{j+1} \leq n} Q_{s_1} V_A Q_{s_2} V_A \cdots V_A Q_{s_{j+1}} C$$

$$\times \sum_{1 \leq m_1 < m_2 < ... < m_{k+1} \leq \tilde{n}} \tilde{Q}_{m_1} V_B \tilde{Q}_{m_2} V_B ... V_B \tilde{Q}_{m_{k+1}}$$

$$\times \hat{I}(s_1, \ldots, s_{j+1}, m_1, \ldots, m_{k+1}),$$

where

$$\hat{I}(s_1, \ldots, s_{j+1}, m_1, \ldots m_{k+1}) := \frac{(-1)^{k+j+1}}{4\pi^2}$$

$$\times \int_{\Gamma_A} \int_{\Gamma_B} \frac{f(z,w)dz\, dw}{(\lambda_{s_1}(A) - z) \cdots (\lambda_{s_{k+1}}(A) - z)(\lambda_{m_1}(B) - w) \cdots (\lambda_{m_{k+1}}(B) - w)}.$$

Let $|M_{jk}|$ be the matrix whose entries are defined by

$$(|M_{jk}|\tilde{e}_t, e_s) = |(M_{jk}\tilde{e}_t, e_s)| \quad (t = 1, ..., \tilde{n}; s = 1, ..., n).$$

In the entry-wise sense, we have

$$|M_{jk}| \leq J_{jk} \sum_{1 \leq s_1 < s_2 < ... < s_{j+1} \leq n} Q_{s_1} |V_A|_e Q_{s_2} |V_A|_e \cdots |V_A|_e Q_{s_{j+1}} |C|$$

$$\times \sum_{1 \leq m_1 < m_2 < ... < m_{k+1} \leq \tilde{n}} \tilde{Q}_{m_1} |V_B|_{\tilde{e}} \tilde{Q}_{m_2} |V_B|_{\tilde{e}} ... |V_B|_{\tilde{e}} \tilde{Q}_{m_{k+1}}$$

where

$$J_{jk} := \max_{1 \leq s_1 < ... < s_{j+1} \leq n; 1 \leq m_1 < ... < m_{k+1} \leq \tilde{n}} |\hat{I}(s_1, \ldots, s_{j+1}, m_1, \ldots m_{k+1})|.$$

But

$$\sum_{1 \leq s_1 < s_2 < ... < s_{j+1} \leq n} Q_{s_1} |V_A|_e Q_{s_2} |V_A|_e \cdots |V_A|_e Q_{s_{j+1}} = |V_A|_e^j$$

and

$$\sum_{1\leq m_1<m_2<...<m_{k+1}\leq\tilde{n}} \tilde{Q}_{m_1}|V_B|_{\tilde{e}}\tilde{Q}_{m_2}|V_B|_{\tilde{e}}\cdots|V_B|_{\tilde{e}}Q_{m_{k+1}} = |V_B|_{\tilde{e}}^k.$$

Thus

$$|M_{jk}| \leq J_{jk}|V_A|_e^j|C||V_B|_{\tilde{e}}^k$$

and therefore,

$$\|M_{jk}\| \leq \||M_{jk}|\| \leq J_{jk}\||V_A|_e^j\|\|C\|\||V_B|_{\tilde{e}}^k\| \quad (j<n, k<\tilde{n}). \tag{5.4}$$

Due to Lemma 4.2

$$J_{jk} \leq \sup_{z\in co(A),w\in co(B)} \frac{|f^{(j,k)}(z,w)|}{j!k!} = \sqrt{j!k!}\psi_{jk} \quad (j,k\geq 0).$$

This inequality, (5.4) and (5.3) imply the required result. Q. E. D.

Lemma 3.4 implies

$$\|V_A^k\| \leq \frac{1}{\sqrt{k!}}N_2^k(V_A). \tag{5.5}$$

Take into account that $N_2(|V_A|_e) = N_2(V_A)$. Thus

$$\||V_A|_e^k\| \leq \frac{1}{\sqrt{k!}}N_2^k(V_A) \; (k=1,...,n-1).$$

The similar inequality holds for V_B. In addition,

$$N_2^2(|C|) = \sum_{j=1}^n \||C|\tilde{e}_j\|^2 = \sum_{j=1}^n\sum_{k=1}^n |(C\tilde{e}_j,e_k)|^2 = \sum_{j=1}^n\sum_{k=1}^n \|C\tilde{e}_j\|^2 = N_2^2(C).$$

Now the previous lemma yields

Lemma 4.4. *Let* $A \in \mathbf{C}^{n\times n}$ *and* $B \in \mathbf{C}^{\tilde{n}\times\tilde{n}}$ *whose nilpotent parts are* V_A *and* V_B *respectively. If, in addition,* $f(z,w)$ *is regular on a neighborhood of* $co(A) \times co(B)$, *then*

$$\|F(f,A,C,B)\| \leq N_2(C)\sum_{j=0}^{n-1}\sum_{k=0}^{\tilde{n}-1} \psi_{jk}N_2^j(V_A)N_2^k(V_B).$$

If $V_A = 0$, $V_B \neq 0$, *and* $f(z,w)$ *is regular on a neighborhood of* $\sigma(A)\times co(B)$, *then*

$$\|F(f,A,C,B)\| \leq N_2(C)\sum_{j=0}^{\tilde{n}-1} \psi_{0j}N_2^j(V_B).$$

If both A *and* B *are normal and* $f(z,w)$ *is regular on a neighborhood of* $\sigma(A) \times \sigma(B)$, *then*

$$\|F(f,A,C,B)\| \leq N_2(C)\sup_{z\in\sigma(A),w\in\sigma(B)} |f(z,w)|.$$

Proof of Theorem 4.3: The previous lemma and the equality $g(A) = N_2(V_A)$ imply the required result. Q. E. D.

4.6   Generalized polynomial matrix equations

As it is shown in Section 2.1, functions defined by (4.1) play an essential role in the theory of the generalized polynomial operator equations

$$\sum_{j=0}^{m_1}\sum_{k=0}^{m_2} c_{jk} A^j X B^k = C \quad (m_1, m_2 < \infty), \qquad (6.1)$$

where c_{jk} are complex numbers; $A \in \mathbf{C}^{n \times n}, B \in \mathbf{C}^{\tilde{n} \times \tilde{n}}$ and $C \in \mathbf{C}^{n \times \tilde{n}}$ again. Let

$$P(z, w) = \sum_{j=0}^{m_1}\sum_{k=0}^{m_2} c_{jk} z^j \tilde{w}^k \neq 0 \quad ((z, w) \in \sigma(A) \times \sigma(B)). \qquad (6.2)$$

Then by Theorem 2.2 equation (6.1) has unique solution which is presentable by the formula

$$X = F\left(\frac{1}{P(z, w)}, A, C, B\right).$$

To apply the latter theorem we should impose the condition stronger than (6.2):

$$P(z, w) \neq 0 \quad ((z, w) \in co(A) \times co(B)). \qquad (6.3)$$

Put

$$\psi_{00}(P) = \sup_{z \in \sigma(A), w \in \sigma(B)} \frac{1}{|P(z, w)|},$$

$$\psi_{jk}(P) = \frac{1}{(j!k!)^{3/2}} \sup_{z \in co(A), w \in co(B)} \left| \frac{\partial^{j+k}}{\partial z^j \partial w^k}\left(\frac{1}{P(z, w)}\right)\right|, \qquad (6.4a)$$

$$\psi_{0k}(P) := \frac{1}{(k!)^{3/2}} \sup_{z \in \sigma(A), w \in co(B)} \left| \frac{\partial^k}{\partial w^k}\left(\frac{1}{P(z, w)}\right)\right|, \qquad (6.4b)$$

and

$$\psi_{j0}(P) := \frac{1}{(j!)^{3/2}} \sup_{z \in co(A), w \in \sigma(B)} \left| \frac{\partial^j}{\partial z^j}\left(\frac{1}{P(z, w)}\right)\right| \quad (j, k \geq 1). \qquad (6.4c)$$

Then Theorem 4.3 implies

Corollary 4.5. *Let condition (6.3) hold. Then the unique solution X of (6.1) is subject to the inequality*

$$\|X\| \leq N_2(C) \sum_{j=0}^{n-1}\sum_{k=0}^{\tilde{n}-1} \psi_{jk}(P) g^j(A) g^k(B).$$

If, in addition, A is normal, then

$$\|X\| \leq N_2(C) \sum_{j=0}^{\tilde{n}-1} \psi_{0j}(P) g^j(B).$$

If both A and B are normal and condition (6.3) holds, then

$$\|X\| \leq N_2(C) \sup_{z \in \sigma(A), w \in \sigma(B)} |f(z,w)|.$$

For example, consider the Sylvester equation

$$AX - XB = C \tag{6.5}$$

assuming that

$$\delta(A, B) := \text{dist } (co(A), co(B)) > 0.$$

With $P(z, w) = z - w$, according to (6.4) we have

$$\psi_{jk}(P) \leq \frac{(k+j)!}{\delta^{j+k+1}(A,B)(k!j!)^{3/2}} \quad (j = 0, 1, ..., n-1; k = 0, 1, ..., \tilde{n}-1).$$

Hence, by Corollary 4.5 a solution of (6.5) satisfies the inequality

$$\|X\| \leq N_2(C) \sum_{k=0}^{n-1} \sum_{j=0}^{\tilde{n}-1} \frac{(k+j)!}{\delta^{j+k+1}(A,B)(k!j!)^{3/2}} g^j(A) g^k(B).$$

4.7 Comments to Chapter 4

The material in this chapter is based on the papers [37, 44]. About applications of the results presented in this chapter to linear and nonlinear non-autonomous systems see the papers [46] and [45], respectively.

Chapter 5

Two-sided Matrix Sylvester Equations

In this chapter we consider the two-sided Sylvester equation

$$\sum_{k=1}^{m} A_{1k} X A_{2k} = C \ (1 \leq m < \infty),$$

where A_{lk} are $n_l \times n_l$ matrices $(k = 1, ..., m; \ l = 1, 2)$, and C is an $n_1 \times n_2$-matrix. We obtain solution estimates for that equation and explore perturbations of solutions. The obtained solution estimates enable to derive the bound for the distance between invariant subspaces of two matrices.

The main tool in this chapter is the norm estimate for the resolvent of the operator

$$\sum_{k=1}^{m} A_{1k}^T \otimes A_{2k},$$

where $\otimes$ means the tensor product and A^T is the matrix transpose of A.

5.1 Auxiliary results

In this chapter $E_1 = \mathbf{C}^{n_1}$ and $E_2 = \mathbf{C}^{n_2}$ are the Euclidean spaces of the dimensions n_1 and n_2, with scalar products $< ., . >_1$ and $< ., . >_2$, respectively, and the norms $\|.\|_l = \sqrt{< ., . >_l} \ (l = 1, 2)$.

The tensor product $H = E_1 \otimes E_2$ of E_1 and E_2 is defined by the following way. Consider the collection of all formal finite sums of the form

$$u = \sum_j y_j \otimes h_j \ (y_j \in E_1, h_j \in E_2)$$

with the understanding that

$$\lambda(y \otimes h) = (\lambda y) \otimes h = y \otimes (\lambda h), (y + y_1) \otimes h = y \otimes h + y_1 \otimes h,$$

$$y \otimes (h + h_1) = y \otimes h + y \otimes h_1 \quad (y, y_1 \in E_1; h, h_1 \in E_2; \lambda \in \mathbf{C}).$$

On that collection define the scalar product as

$$< y \otimes h, y_1 \otimes h_1 >_H = < y, y_1 >_1 < h, h_1 >_2$$

and take the norm $\|.\| = \|.\|_H = \sqrt{< ., . >_H}$. In addition, I_H and I_l mean the unit operators in H and E_l, respectively; sometimes we omit the indexes l and H.

Furthermore, $\mathcal{B}(E)$ denotes the set of all linear operators in a space E. So $\mathcal{B}(E_l) = \mathbf{C}^{n_l \times n_l}$, where $\mathbf{C}^{n_1 \times n_2}$ means the set of all complex $n_1 \times n_2$ matrices.

The Kronecker product of $A \in \mathcal{B}(E_1)$ and $B \in \mathcal{B}(E_2)$ denoted by $A \otimes B$ is defined by

$$(A \otimes B)(f_1 \otimes f_2) = (Af_1) \otimes (Bf_2) \quad (f_l \in \mathbf{C}^{n_l}, l = 1, 2).$$

In the equivalent form the tensor product of $A = (a_{ij}) \in \mathbf{C}^{n_1 \times n_1}$ and B is defined to be the block matrix

$$A \otimes B = \begin{pmatrix} a_{11}B & \cdots & a_{1n_1}B \\ \cdot & \cdots & \cdot \\ \cdot & \cdots & \cdot \\ \cdot & \cdots & \cdot \\ a_{n_11}B & \cdots & a_{n_1n_1}B \end{pmatrix},$$

cf. [59].

Some very basic properties of the Kronecker product are stated in the following lemma.

Lemma 5.1. *With obvious notations, we have*

(a) $(A + A_1) \otimes B = A \otimes B + A_1 \otimes B; A \otimes (B_1 + B_2) = A \otimes B_1 + A \otimes B_2;$

$(\lambda A \otimes B) = \lambda (A \otimes B); \quad (A \otimes \lambda B) = \lambda (A \otimes B) \quad (\lambda \in \mathbf{C}).$

(b) $I_1 \otimes I_2 = I_H.$

(c) $(A \otimes B)(C \otimes D) = (AC) \otimes (BD).$

(d) $(A \otimes B)^* = A^* \otimes B^*.$

(e) $\|A \otimes B\| = \|A\|\|B\|.$

(f) $A \otimes B$ *is invertible if and only if A and B are both invertible, in which case* $(A \otimes B)^{-1} = A^{-1} \otimes B^{-1}.$

Proof. (a) For example,

$$((A + A_1) \otimes B)(x \otimes y) = (Ax + A_1x) \otimes By = Ax \otimes By + A_1x \otimes By =$$

$$(A \otimes B)(x \otimes y) + (A_1 \otimes B)(x \otimes y) \quad (x \in E_1, y \in E_2).$$

Similarly the other relations from (a) can be proved.

(b) $(I \times I)(x \otimes y) = x \otimes y = I(x \otimes y).$

(c) $(A \otimes B)(C \otimes D)(x \otimes y) = (A \otimes B)(Cx \otimes Dy) = (ACx \otimes BDy) =$

$$[(AC) \otimes (BD)](x \otimes y).$$

(d) $< (A \otimes B)^*(x \otimes y), u \otimes v > = < x \otimes y, (Au \otimes Bv) >$

$$= < x, Au >_1 < y, Bv >_2 =$$

$$< A^*x, u >_1 < B^*y, v >_2$$

$$= < A^*x \otimes B^*y, u \otimes y > = < (A^* \otimes B^*)(x \otimes y), u \otimes y > .$$

(e) Choose $x \in \mathbf{C}^{n_1}$, $y \in \mathbf{C}^{n_2}$, $\|x\| = \|y\| = 1$ and $\|Ax\| = \|A\|, \|By\| = \|B\|$. Then

$$\|(A \otimes B)(x \otimes y)\| = \|Ax \otimes By\| = \|Ax\|\|By\| = \|A\|\|B\|.$$

(f) If A and B are invertible, then $(A \otimes B)(A^{-1} \otimes B^{-1}) = AA^{-1} \otimes BB^{-1} = I$ and similarly $(A^{-1} \otimes B^{-1})(A \otimes B) = I$. Conversely, suppose $A \otimes B$ is invertible. Since

$$A \otimes B = (A \otimes I)(I \otimes B),$$

it follows that $A \otimes I$ and $I \otimes B$ are also invertible, so it will suffice to show that the invertibility of $A \otimes I$ implies that of A (the proof for B is similar). We know that $A \otimes I$ is bounded from below, and it will suffice to show that A is bounded from below. Thus we are reduced to showing that the boundedness from below of $A \otimes I$ implies that of A. By supposition, there exists an $\epsilon > 0$ such that $\|(A \otimes I)u\| \geq \epsilon\|u\|$ for all $u \in H$. Then $\|(A \otimes I)x \otimes y\| \geq \epsilon\|x\|\|y\|$ for all $x \in \mathbf{C}^{n_1}, y \in \mathbf{C}^{n_2}$, that is, $\|Ax \otimes y\| \geq \epsilon\|x\|\|y\|$ whence $\|Ax\| \geq \epsilon\|x\|$ (choose any nonzero y, then cancel). Q. E. D.

Corollary 5.1. *If $A \in \mathcal{B}(E_1)$ and $B \in \mathcal{B}(E_2)$ are either (a) unitary, (b) selfadjoint, (c) positive definite, (d) normal then so is $A \otimes B$.*

Indeed, for example, let A and B be unitary. Then $(A \otimes B)^*(A \otimes B) = A^*A \otimes B^*B = I$ and similarly $(A \otimes B)(A \otimes B)^* = I$.

Similarly the other assertions can be proved.

Lemma 5.2. *The spectrum of* $A \otimes B$ *is*

$$\sigma(A \otimes B) = \{ts : t \in \sigma(A), s \in \sigma(B)\}.$$

For the proof see [59].

From the latter theorem it follows

$$\text{trace } (A \otimes B) = \text{trace } (A) \text{ trace } (B).$$

Since

$$(A \otimes B)^*(A \otimes B) = A^*A \otimes B^*B$$

and

$$[(A \otimes B)^*(A \otimes B)]^p = (A^*A)^p \otimes (B^*B)^p \ (p > 0),$$

we arrive at the relation

$$\text{trace } [(A \otimes B)^*(A \otimes B)]^p = \text{trace } (A^*A)^p \text{ trace } (B^*B)^p.$$

We thus obtain

Corollary 5.2. *One has* $N_p(A \otimes B) = N_p(A)N_p(B) \ (p \geq 1)$.

Recall that $N_p(A) = \sqrt[p]{\text{trace } (A^*A)^{p/2}}$.

Our main object in this chapter is the operator Z defined on H by

$$Z = \sum_{k=1}^{m} A_{1k} \otimes A_{2k} \ (1 \leq m < \infty), \tag{1.1}$$

where A_{lk} are $n_l \times n_l$ matrices $(k = 1, ..., m; \ l = 1, 2)$.

5.2   The resolvent of operator Z

5.2.1   *Statement of the result*

Let $\{e_k\}_{k=1}^n$ be some Schur basis of $A \in \mathbf{C}^{n \times n}$. So $D_A e_j = \lambda_j(A)e_j \ (j = 1, ..., n)$ and

$$V_A e_k = a_{1k}e_1 + \cdots + a_{k-1,k}e_{k-1} \ (k = 2, ..., n), V_A e_1 = 0$$

are the diagonal part and nilpotent one of A, respectively (see Section 3.1).

In addition, $|A| = |A|_e$ means the operator, whose entries in basis $\{e_k\}$ are the absolute values of the entries of operator A in that basis. We will call $|A|$ the absolute value of A (with respect to basis $\{e_k\}$).

The smallest integer $\nu_A \leq n$, such that $|V_A|_e^{\nu_A} = 0$ will be called the nilpotency index of A.

Note that in the general case the nilpotency index ν_Z of Z satisfies the inequality $\nu_Z \leq n_1 n_2$.

Recall that

$$g(A) = (N_2^2(A) - \sum_{k=1}^{n} |\lambda_k(A)|^2)^{1/2} \quad (A \in \mathbf{C}^{n \times n})$$

and

$$g^2(A) \leq N_2^2(A - A^*)/2 \tag{2.1}$$

(see Section 3.1). Note that

$$Z - Z^* = \sum_{k=1}^{m} [(A_{1k} - A_{1k}^*) \otimes A_{2k} + A_{1k}^* \otimes (A_{2k} - A_{2k}^*)].$$

Thus, due to (2.1)

$$g(Z) \leq \frac{1}{\sqrt{2}} N_2(Z - Z^*) \leq g_0(Z), \tag{2.2}$$

where

$$g_0(Z) := \frac{1}{\sqrt{2}} \sum_{k=1}^{m} [N_2(A_{1k} - A_{1k}^*)N_2(A_{2k}) + N_2(A_{2k} - A_{2k}^*)N_2(A_{1k})].$$

Theorem 5.1. *Let Z be defined by (1.1). Then for all $\lambda \notin \sigma(Z)$ one has*

$$\|(Z - \lambda I_H)^{-1}\| \leq \sum_{j=0}^{\nu_Z - 1} \frac{g_0^j(Z)}{\sqrt{j!} \rho^{j+1}(Z, \lambda)}.$$

This theorem is proved in the next subsection. Besides here $g_0^0(Z) = 1$. Recall that $\|.\|$ means the spectral norm.

If A_{lk} are Hermitian for all $l = 1, 2; k = 1, ..., m$, then $g_0(Z) = 0$ and Theorem 5.1 implies the inequality

$$\|(Z - \lambda I_H)^{-1}\| \leq \frac{1}{\rho(Z, \lambda)},$$

which in fact is the equality. So Theorem 5.1 is sharp.

5.2.2   *Proof of Theorem 5.1*

We write $A \geq 0$, if all the entries of A in the Schur basis are nonnegative, and $A \geq B$ if $A - B \geq 0$.

Lemma 5.3. *Let A be an $n \times n$ matrix. Then*

$$|(A - \lambda I)^{-1}|_e \leq \sum_{j=0}^{\nu_A - 1} \frac{1}{\rho^{j+1}(A, \lambda)} |V_A|_e^j \quad (\lambda \notin \sigma(A)),$$

where $\rho(A, \lambda) = \min_{k=1,\dots,n} |\lambda - \lambda_k(A)|$ and V_A is the nilpotent part of A.

Proof. Due to the triangular representation,

$$A - \lambda I = D_A + V_A - \lambda I = (D_A - \lambda I)(I - Q_\lambda),$$

where $Q_\lambda = -(D_A - \lambda I)^{-1} V_A$ is nilpotent, since V_A in Schur's basis is a nilpotent triangular matrix and $(D_A - \lambda I)^{-1}$ is a diagonal one. Hence,

$$(A - \lambda I)^{-1} = (I - Q_\lambda)^{-1} (D_A - \lambda I)^{-1} \quad (\lambda \notin \sigma(D_A)).$$

But $\sigma(A) = \sigma(D_A)$ and $|(D_A - \lambda I)^{-1}|_e \leq \frac{1}{\rho(D_A, \lambda)} I$ and therefore

$$|Q_\lambda| \leq \frac{1}{\rho(D_A, \lambda)} |V_A|.$$

So $Q_\lambda^{\nu_A} = 0$ and thus,

$$(A - \lambda I)^{-1} = (D_A - \lambda I)^{-1} \sum_{k=0}^{\nu_A - 1} Q_\lambda^k \quad (\lambda \notin \sigma(A)). \tag{2.3}$$

This proves the required result. Q. E. D.

Lemma 5.4. *One has*

$$\|(A - \lambda I)^{-1}\| \leq \sum_{j=0}^{\nu_A - 1} \frac{g^j(A)}{\sqrt{j!} \rho^{j+1}(A, \lambda)} \quad (\lambda \notin \sigma(A)).$$

Proof. Due to Lemma 3.4,

$$\| |V_A|^j \| \leq \frac{N_2^j(|V_A|)}{\sqrt{j!}} \quad (j = 1, \dots, \nu_A - 1). \tag{2.4}$$

In addition, due to Lemma 3.1, $N_2(|V_A|) = N_2(V_A) = g(A)$. Since $\|A\| \leq \| |A| \|$, the previous lemma yields the required result. Q. E. D.

The later lemma is a slight refinement of Theorem 3.2.

The assertion of Theorem 5.1 is due to Lemma 5.4 and inequality (2.2). Q. E. D.

5.3 Simultaneously triangularizable matrices

In this section we considerably improve Theorem 5.1 in a particular but practically important case.

A family of operators on $\mathbf{C}^n$ is said to be simultaneously triangularizable, if they can be reduced to the triangular form by the same unitary operator. That is, they have a joint Schur basis and therefore all of their invariant subspaces are joint.

Throughout this section it is supposed that operators

$$A_{1j} \ (j = 1, ..., m) \text{ are simultaneously triangularizable} \qquad (3.1a)$$

as well as operators

$$A_{2j} \ (j = 1, ..., m) \text{ are simultaneously triangularizable.} \qquad (3.1b)$$

The important examples of operators under condition (3.1) are the Kronecker's sum $A_1 \otimes I + I \otimes A_2$ and Kronecker product $A_1 \otimes A_2$. Simple calculations show that the eigenvalues of Z are

$$\lambda_{st}(Z) = \sum_{k=1}^{m} \lambda_s(A_{1k})\lambda_t(A_{2k}) \ (s = 1, ..., n_1; \ t = 1, ..., n_2), \qquad (3.2)$$

provided conditions (3.1) hold. We need the triangular representations of A_{lk}:

$$A_{lk} = D_{lk} + V_{lk} \ (\sigma(A_{lk}) = \sigma(D_{lk}); \ k = 1, ..., m; l = 1, 2), \qquad (3.3)$$

where D_{lk} and V_{lk} are the diagonal and nilpotent parts of A_{lk}, respectively. From the definition of Z it follows that

$$Z = D_Z + V_Z \text{ with } D_Z = \sum_{k=1}^{m} D_{1k} \otimes D_{2k} \qquad (3.4)$$

and

$$V_Z = \sum_{k=1}^{m}(D_{1k} \otimes V_{2k} + V_{1k} \otimes D_{2k} + V_{1k} \otimes V_{2k}). \qquad (3.5)$$

To formulate our next result put

$$\hat{r}_l = \max_{k=1,...,m} r_s(A_{lk}), \hat{g}_l = \sum_{k=1}^{m} g(A_{lk}) \ (l = 1, 2)$$

and

$$\eta_{k_1,k_2}^{(j)} = \frac{j!}{(k_1!k_2!)^{3/2}[(j - k_1 - k_2)!]^2}.$$

Theorem 5.2. *Let Z be defined by (1.1) and conditions (3.1) hold. Then*

$$\nu_Z \leq n_1 + n_2 + \min\{n_1, n_2\} - 2 \tag{3.6}$$

and

$$\|(Z - \lambda I_H)^{-1}\| \leq$$

$$\sum_{j=0}^{\nu_Z - 1} \frac{1}{\rho^{j+1}(Z, \lambda)} \sum_{0 \leq k_1 + k_2 \leq j} \eta_{k_1, k_2}^{(j)} \hat{r}_1^{k_1} \hat{r}_2^{k_2} \hat{g}_1^{j-k_1} \hat{g}_2^{j-k_2} \quad (\lambda \notin \sigma(Z)) \tag{3.7}$$

with

$$\rho(Z, \lambda) = \min_{s=1,\ldots,n_1; \, t=1,\ldots,n_2} \left| \lambda - \sum_{k=1}^{m} \lambda_s(A_{1k}) \lambda_t(A_{2k}) \right|.$$

The proof of this theorem is presented in the next section. Note that $1/(j-k)! = 0$ for $j < k$.

If A_{lk} are normal for all $l = 1, 2; k = 1, \ldots, m$, then $\hat{g}_1 = \hat{g}_2 = 0$. Besides, $\hat{g}_1^0 = \hat{g}_2^0 = 1$. Since

$$\eta_{j,j}^{(j)} = \frac{1}{(j!)^2[(-j)!]^2} = 0 \ (j > 0), \eta_{0,0}^{(0)} = 1,$$

(3.7) implies the inequality $\|(Z - \lambda I_H)^{-1}\| \leq \frac{1}{\rho(Z,\lambda)}$. But Z is normal in this case, therefore the latter inequality becomes equality. So Theorem 5.2 is sharp.

In addition, we will show that Theorem 5.2 implies

Corollary 5.3. *Let conditions (3.1) hold. Then*

$$\|(Z - \lambda I_H)^{-1}\| \leq \sum_{j=0}^{\nu_Z - 1} \frac{3^{j/2}}{\sqrt{j!} \rho^{j+1}(Z, \lambda)} (\hat{r}_2 \hat{g}_1 + \hat{r}_1 \hat{g}_2 + \hat{g}_1 \hat{g}_2)^j \quad (\lambda \notin \sigma(Z)).$$

5.4　Proof of Theorem 5.2

To check inequality (3.6) we need the following lemma.

Lemma 5.5. *Let W_k $(k = 1, 2, 3)$ be mutually commuting nilpotent operators in $\mathbf{C}^n$, and $W_k^{j_k} = 0$ for an integer $j_k \leq n$. Then*

$$(W_1 + W_2)^{j_1 + j_2 - 1} = 0 \tag{4.1}$$

and

$$(W_1 + W_2 + W_3)^{j_1 + j_2 + j_3 - 2} = 0. \tag{4.2}$$

Proof. Put $\hat{W}_2 = W_1 + W_2$. We have

$$\hat{W}_2^{j_1+j_2-1} = \sum_{k=0}^{j_1+j_2-1} \binom{j_1+j_2-1}{k} W_1^{j_1+j_2-k-1} W_2^k,$$

where $\binom{j}{k}$ are the binomial coefficients. Obviously, either $j_1+j_2-k-1 \geq j_1$, or $k \geq j_2$. This proves (4.1). Furthermore, put $\hat{W}_3 = W_1 + W_2 + W_3 = \hat{W}_2 + W_3$. Since $\hat{W}_2^{j_1+j_2-1} = 0$, replacing in our arguments W_1 by $\hat{W}_2$, and W_2 by W_3 we obtain $\hat{W}_3^{j_1+j_2+j_3-2} = 0$, as claimed. Q. E. D.

Let $V_l, B_l \in \mathbf{C}^{n_l \times n_l}$ $(l = 1, 2)$. Besides, V_l are nilpotent, B_l are normal. In addition, V_1 and B_1 have joint invariant subspaces, as well as V_2 and B_2 have joint invariant subspaces. Consider the operator

$$W = V_1 \otimes B_2 + B_1 \otimes V_2 + V_1 \otimes V_2. \tag{4.3}$$

Then the Schur basis of W is defined as $\{e_j^{(1)} \otimes e_k^{(2)}\}_{j=1,\ldots,n_1; k=1,\ldots,n_2}$, where $\{e_j^{(l)}\}_{j=1}^{n_l}$ is the joint Schur basis of V_l and B_l.
So if

$$W(e_j^{(1)} \otimes e_k^{(2)}) = \sum_{j_1=1}^{j} \sum_{k_1=1}^{k} w_{j_1 j k_1 k}(e_{j_1}^{(1)} \otimes e_{k_1}^{(2)}),$$

then

$$|W|(e_j^{(1)} \otimes e_k^{(2)}) = \sum_{j_1=1}^{j} \sum_{k_1=1}^{k} |w_{j_1 j k_1 k}|(e_{j_1}^{(1)} \otimes e_{k_1}^{(2)}).$$

Lemma 5.6. *Let W be defined by (4.3) and $\nu = n_1 + n_2 + \min\{n_1, n_2\} - 2$. Then $W^\nu = 0$ and*

$$|W|^j \leq$$

$$\sum_{k_1+k_2+k_3=j} C_{k_1,k_2,k_3}^j r_s^{k_1}(B_2) r_s^{k_2}(B_1)(|V_1|^{k_1} \otimes I)(I \otimes |V_2|^{k_2})(|V_1|^{k_3} \otimes |V_2|^{k_3})$$

$(j < \nu)$, where

$$C_{k_1,k_2,k_3}^j = \frac{j!}{k_1! k_2! k_3!}.$$

Proof. In the Schur basis of W we have

$$|W| \leq r_s(B_1)I \otimes |V_2| + |V_1| \otimes r_s(B_2)I + |V_1| \otimes |V_2|.$$

Put $W_1 = r_s(B_1)I \otimes |V_2|, W_2 = |V_1| \otimes r_s(B_2)I, W_3 = |V_1| \otimes |V_2|$. Since they commute and $|V_l|^{n_l} = 0$, due to (3.10) we have $|W|^\nu = 0$. In addition, we can write

$$(W_1 + W_2 + W_3)^j = \sum_{k_1+k_2+k_3=j} C_{k_1,k_2,k_3}^j W_1^{k_1} W_2^{k_2} W_3^{k_3}.$$

This proves the lemma. Q. E. D.

Lemma 5.7. *Let conditions (3.1) hold. Then inequality (3.6) is valid.*

Proof. From (3.5) it follows

$$|V_Z| \leq \sum_{k=1}^m (r_s(D_{2k})|V_{1k}| \otimes I + r_s(D_{1k})I \otimes |V_{2k}| + |V_{1k}| \otimes |V_{2k}|).$$

Hence with the notations

$$\hat{V}_l = |V_{lk}| + \cdots + |V_{lm}|$$

and

$$\hat{r}_l = \max_{k=1,\dots,m} r_s(D_{lk}) = \max_{k=1,\dots,m} r_s(A_{lk}) \ (l=1,2),$$

we get

$$|V_Z| \leq \hat{r}_2\hat{V}_1 \otimes I + \hat{r}_1 I \otimes \hat{V}_2 + \hat{V}_1 \otimes \hat{V}_2 = W_1 + W_2 + W_3, \tag{4.4}$$

where $W_1 = \hat{r}_2\hat{V}_1 \otimes I, W_2 = \hat{r}_1 I \otimes \hat{V}_2, W_3 = \hat{V}_1 \otimes \hat{V}_2$. Observe that W_1, W_2, and W_3 are mutually commuting nilpotent operators. Applying the previous lemma, we have $|V_Z|^{n_1+n_2+\min\{n_1,n_2\}-2} = 0$, as claimed. Q. E. D.

Proof of Theorem 5.2: From Lemma 5.3 it follows

$$\|(Z - \lambda I)^{-1}\| \leq \sum_{j=0}^{\nu_Z-1} \frac{1}{\rho^{j+1}(Z,\lambda)} \||V_Z|^j\|. \tag{4.5}$$

According to (4.4) and Lemma 5.6,

$$|V_Z|^j \leq \sum_{k_1+k_2+k_3=j} C_{k_1,k_2,k_3}^j \hat{r}_2^{k_1}(\hat{V}_1^{k_1} \otimes I)\hat{r}_1^{k_2}(I \otimes \hat{V}_2^{k_2})(\hat{V}_1^{k_3} \otimes \hat{V}_2^{k_3}) =$$

$$\sum_{k_1+k_2+k_3=j} C_{k_1,k_2,k_3}^j \hat{r}_2^{k_1}(\hat{V}_1^{k_1+k_3} \otimes I)\hat{r}_1^{k_2}(I \otimes \hat{V}_2^{k_2+k_3}) \ (j < \nu).$$

Now the relations $g(A_{lk}) = N_2(V_{lk})$ $(l = 1, 2; k = 1, ..., m)$ and

$$N_2(V_l) \le N_2(V_{l1}) + ... + N_2(V_{lm}) = \hat{g}_l$$

imply

$$N_2^j(V_Z) \le \sum_{k_1+k_2+k_3=j} C_{k_1,k_2,k_3}^j \hat{r}_2^{k_1} \hat{r}_1^{k_2} \frac{\hat{g}_1^{k_1+k_3} \hat{g}_2^{k_2+k_3}}{\sqrt{k_1!k_2!k_3!}}. \quad (4.6)$$

Taking into account that $k_3 = j - k_1 - k_2$, we have

$$\||V_Z|^j\| \le \sum_{0 \le k_1+k_2 \le j} C_{k_1,k_2,j-k_1-k_2}^j \hat{r}_2^{k_1} \hat{r}_1^{k_2} \frac{\hat{g}_1^{j-k_2} \hat{g}_2^{j-k_1}}{\sqrt{k_1!k_2!(j-k_1-k_2)!}}$$

Or

$$\||V_Z|^j\| \le \sum_{0 \le k_1+k_2 \le j} \eta_{k_1,k_2}^{(j)} \hat{r}_2^{k_1} \hat{r}_1^{k_2} \hat{g}_1^{j-k_2} \hat{g}_2^{j-k_1}. \quad (4.7)$$

This and (4.5) prove the theorem. Q. E. D.

Proof of Corollary 5.3: For $k_1 + k_2 + k_3 = j$ we have

$$\frac{1}{k_1!k_2!(k_3!)^2} \le \frac{1}{k_1!k_2!k_3!} = \frac{j!}{j!k_1!k_2!k_3!} = C_{k_1,k_2,k_3}^j \frac{1}{j!} \le \frac{3^j}{j!},$$

since

$$(a + b + c)^j = \sum_{k_1+k_2+k_3=j} C_{k_1,k_2,k_3}^j a^{k_1} b^{k_2} c^{k_3}.$$

Now (4.6) and (4.5) imply the required result. Q. E. D.

5.5   Particular cases of conditions (3.1)

Assuming that conditions (3.1) hold, in this section we improve inequality (3.6) under additional conditions. First suppose that

for each index $k \le m$ at least one of operators A_{1k} or A_{2k} is normal.
(5.1)

Then $V_{1k} \otimes V_{2k} = 0$ $(k = 1, ..., m)$ and therefore from (3.5) it follows,

$$V_Z = \sum_{k=1}^m (V_{1k} \otimes D_{2k} + D_{1k} \otimes V_{2k}). \quad (5.2)$$

Hence,

$$|V_Z| \le \sum_{k=1}^m (r_s(A_{2k})|V_{1k}| \otimes I + r_s(A_{1k})I \otimes |V_{2k}|) \le \hat{r}_2 \hat{V}_1 \otimes I + \hat{r}_1 I \otimes \hat{V}_2,$$

where $\hat{V}_l, \hat{r}_l$ are defined as in the previous section. Since $\hat{V}_l^{n_l} = 0$, and the operators $\hat{V}_1 \otimes I$ and $I \otimes \hat{V}_2$ commute, Lemma 5.5 implies the equality $V_Z^{n_1+n_2-1} = 0$. Moreover,

$$|V_Z|^j \le \sum_{k=0}^{j} \binom{j}{k} \hat{r}_2^{j-k} \hat{r}_1^k (\hat{V}_1^{j-k} \otimes I)(I \otimes \hat{V}_2^k) \quad (j < n_1 + n_2 - 1).$$

Hence,

$$\||V_Z|^j\| \le \sum_{k=0}^{j} \binom{j}{k} \hat{r}_2^{j-k} \hat{r}_1^k (\|\hat{V}_1^{j-k}\| \|\hat{V}_2^k\|) \quad (j < n_1 + n_2 - 1)$$

and according to (2.4),

$$\||V_Z|^j\| \le \sum_{k=0}^{j} \binom{j}{k} \hat{r}_1^k \hat{r}_2^{j-k} \frac{N_2^k(\hat{V}_2) N_2^{j-k}(\hat{V}_1)}{\sqrt{(j-k)!k!}}.$$

But $N_2(\hat{V}_l) \le \hat{g}_l$ and therefore,

$$N_2^j(|V_Z|) \le \sum_{k=0}^{j} \binom{j}{k} \hat{r}_2^{j-k} \hat{r}_1^k \frac{\hat{g}_1^{j-k} \hat{g}_2^k}{\sqrt{(j-k)!k!}}$$

for $j < n_2 + n_1 - 1$. Since $V_Z^{n_1+n_2-1} = 0$, making use Lemma 5.3, we arrive at

Theorem 5.3. *Let Z be defined by (1.1) and conditions (3.1) and (5.1) hold. Then*

$$\|(Z - \lambda I_H)^{-1}\| \le \sum_{j=0}^{n_1+n_2-2} \frac{1}{\rho^{j+1}(Z,\lambda)} \sum_{k=0}^{j} \binom{j}{k} \frac{\hat{r}_1^k \hat{r}_2^{j-k} \hat{g}_2^k \hat{g}_1^{j-k}}{\sqrt{(j-k)!k!}} \quad (\lambda \notin \sigma(Z)).$$

$$(5.3)$$

Note that

$$\frac{1}{(j-i)!i!} \le \frac{1}{j!} \sum_{k=0}^{j} \binom{j}{k} \le \frac{2^j}{j!} \quad . \text{ So } \frac{1}{\sqrt{(j-k)!k!}} \le \frac{2^{j/2}}{\sqrt{j!}} \quad (k \le j).$$

Now from (5.3) it follows

Corollary 5.4. *Under the hypothesis of Theorem 5.3 one has*

$$\|(Z - \lambda I_H)^{-1}\| \le \sum_{j=0}^{n_1+n_2-2} \frac{2^{j/2}}{\sqrt{j!}\rho^{j+1}(Z,\lambda)} (\hat{r}_1 \hat{g}_2 + \hat{r}_2 \hat{g}_1)^j.$$

Furthermore, assume that

$$\text{operators } A_{1k} \text{ are normal for all indexes } k = 1, ..., m; \qquad (5.4)$$

A_{2k} can be arbitrary operators satisfying (3.1b). Then

$$V_Z = \sum_{k=1}^{m} D_{1k} \otimes V_{2k},$$

and with absolute values with respect to the Schur basis,

$$|V_Z| = |V_Z|_e \le \sum_{k=1}^{m} r_s(A_{1k}) I \otimes |V_{2k}| \le \hat{r}_1 I \otimes \hat{V}_2,$$

and consequently, $|V_Z|^j \le \hat{r}_1^j I \otimes \hat{V}_2^j$. Now (2.4) yields the inequality

$$\||V_Z|^j\| \le \hat{r}_1^j \frac{N_2^j(\hat{V}_2)}{\sqrt{j!}} \le \hat{r}_1^j \frac{\hat{g}_2^j}{\sqrt{j!}}.$$

But $V_Z^{n_2} = 0$. Hence, making use (4.5), we arrive at

Theorem 5.4. *Let Z be defined by (1.1). Assume that conditions (3.1) and (5.4) hold. Then*

$$\|(Z - \lambda I_H)^{-1}\| \le \sum_{j=0}^{n_2-1} \frac{\hat{r}_1^j \hat{g}_2^j}{\sqrt{j!}\rho^{j+1}(Z, \lambda)} \quad (\lambda \notin \sigma(Z)).$$

One can exchange the places of A_{1k} and A_{2k}.

Consider the Kronecker sum

$$A_1 \oplus A_2 = A_1 \otimes I + I \otimes A_2 \quad (A_l \in \mathbf{C}^{n_l \times n_l}; \ l = 1, 2).$$

So $V_Z = V_1 \otimes I + I \otimes V_2$ and therefore

$$|V_Z|^j \le \sum_{k=0}^{j} \binom{j}{k}(|V_1|^{j-k} \otimes I)(I \otimes |V_2|^k) \quad (j < n_1 + n_2 - 1).$$

Hence,

$$\||V_Z|^j\| \le \sum_{k=0}^{j} \binom{j}{k}(\||V_1|^{j-k}\| \||V_2|^k\|)$$

and according to (2.4),

$$\||V_Z|^j\| \le \sum_{k=0}^{j} \binom{j}{k} \frac{N_2^k(V_2) N_2^{j-k}(V_1)}{\sqrt{(j-k)!k!}} = \sum_{k=0}^{j} \binom{j}{k} \frac{g^k(A_2) g^{j-k}(A_1)}{\sqrt{(j-k)!k!}}.$$

Take into account that

$$\lambda_{ik}(A_1 \oplus A_2) = \lambda_i(A_1) + \lambda_k(A_2) \quad (i = 1, ..., n_1; \ k = 1, ..., n_2). \tag{5.5}$$

Now Lemma 5.3 implies

Corollary 5.5. *For all* $\lambda \notin \sigma(A_1 \oplus A_2)$ *one has*

$$\|(A_1 \oplus A_2 - \lambda I_H)^{-1}\| \le$$

$$\sum_{j=0}^{n_1+n_2-2} \frac{1}{\rho^{j+1}(A_1 \oplus A_2, \lambda)} \sum_{k=0}^{j} \binom{j}{k} \frac{g^k(A_2)g^{j-k}(A_1)}{\sqrt{(j-k)!k!}}$$

and (according to Corollary 5.4),

$$\|(A_1 \oplus A_2 - \lambda I_H)^{-1}\| \le \sum_{j=0}^{n_1+n_2-2} \frac{2^{j/2}(g(A_2) + g(A_1))^j}{\sqrt{j!}\rho^{j+1}(A_1 \oplus A_2, \lambda)}.$$

If A_1 is normal, then $g(A_1) = 0$ and due to Corollary 5.5,

$$\|(A_1 \oplus A_2 - \lambda I_H)^{-1}\| \le \sum_{j=0}^{n_2-1} \frac{g^j(A_2)}{\sqrt{j!}\rho^{j+1}(A_1 \oplus A_2, \lambda)}. \tag{5.6}$$

Furthermore,

$$\lambda_{jk}(A_1 \otimes A_2) = \lambda_j(A_1)\lambda_k(A_2) \quad (j = 1, ..., n_1; \ k = 1, ..., n_2).$$

From Theorem 5.2 it immediately follows

Corollary 5.6. *For all* $\lambda \notin \sigma(A_1 \otimes A_2)$ *one has*

$$\|(A_1 \otimes A_2 - \lambda I_H)^{-1}\| \le \sum_{j=0}^{n_1+n_2+\min\{n_1,n_2\}-3} \frac{\chi_j(A_1, A_2)}{\rho^{j+1}(A_1 \otimes A_2, \lambda)}$$

where

$$\chi_j(A_1, A_2) := \sum_{0 \le k_1+k_2 \le j} \eta^{(j)}_{k_1,k_2} r^{k_1}_s(A_1) r^{k_2}_s(A_2) g^{j-k_1}(A_1) g^{j-k_2}(A_2)$$

and

$$\rho(A_1 \otimes A_2, \lambda) = \min_{s,t} |\lambda_t(A_1)\lambda_s(A_2) - \lambda|.$$

In particular, if A_1 is normal, then according to Theorem 5.4,

$$\|(A_1 \otimes A_2 - \lambda I_H)^{-1}\| \le \sum_{j=0}^{n_2-1} \frac{r^j_s(A_1)g^j(A_2)}{\sqrt{j!}\rho^{j+1}(A_1 \otimes A_2, \lambda)}. \tag{5.7}$$

5.6 Solution estimates for two-sided matrix Sylvester equations

Consider the equation

$$\sum_{s=1}^{m} A_{1s} X A_{2s} = C, \tag{6.1}$$

where $A_{lk} \in \mathbf{C}^{n_l \times n_l}$ and $C \in \mathbf{C}^{n_1 \times n_2}$ are given; $X \in \mathbf{C}^{n_1 \times n_2}$ should be found. Following [59], with each matrix $A = (a_{ij}) \in \mathbf{C}^{s \times n}$ we associate the vector vec $(A) \in \mathbf{C}^{sn}$ defined by

$$\text{vec } A := \text{column } (a_{11}, ..., a_{s1}, a_{12}, ..., a_{s2}, ..., a_{1n}, ..., a_{sn}).$$

Put

$$K := \sum_{s=1}^{m} A_{2s}^T \otimes A_{1s}, \tag{6.2}$$

where A^T is the transpose of A. Equation (6.1) can be written as $K \text{ vec } (X) = \text{vec } (C)$ [59, p. 255]. Let $r_{\text{low}}(K)$ be the lower spectral radius of K: $r_{\text{low}}(K) := \min_k |\lambda_k(K)|$ and assume that $r_{\text{low}}(K) > 0$. Then K is invertible. So vec $(X) = K^{-1}$ vec (C). Clearly, $\|\text{vec } X\| = N_2(X)$. Hence

$$N_2(X) \leq \|K^{-1}\| N_2(C). \tag{6.3}$$

Due to Theorem 5.1

$$\|K^{-1}\| \leq \gamma(K) := \sum_{j=0}^{\nu_K - 1} \frac{g^j(K)}{\sqrt{j!} r_{\text{low}}^{j+1}(K)} \tag{6.4}$$

with $\nu_K \leq n_1 n_2$. We thus have proved

Corollary 5.7. *Let $r_{\text{low}}(K) > 0$. Then (6.1) has a unique solution X and $N_2(X) \leq \gamma(K) N_2(C)$.*

Similarly Theorems 5.2, 5.3 and 5.4 together with inequality (6.3) enable us to estimate solutions of equation (6.1) under conditions (3.1), (5.1) and (5.4).

For example, consider the Sylvester equation

$$AX - XB = C, \tag{6.5}$$

where $A \in \mathbf{C}^{n_1 \times n_1}, B \in \mathbf{C}^{n_2 \times n_2}$. In this case $r_{\text{low}}(K) = \rho_S(A, B) := \min_{j,k} |\lambda_k(A) - \lambda_j(B)|$, and due to Corollary 5.5 we can write

$$\|K^{-1}\| \leq \gamma_S(A, B) := \sum_{j=0}^{n_1 + n_2 - 2} \frac{j!}{\rho_S^{j+1}(A, B)} \sum_{k=0}^{j} \frac{g^k(B) g^{j-k}(A)}{((j-k)! k!)^{3/2}} \tag{6.6}$$

and

$$\|K^{-1}\| \le \hat{\gamma}_S(A,B) := \sum_{j=0}^{n_1+n_2-2} \frac{2^{j/2}}{\sqrt{j!}\rho_S^{j+1}(A,B)} (g(A)+g(B))^j. \qquad (6.7)$$

According to inequality (6.3) we get

Corollary 5.8. *Let condition $\rho_S(A,B) > 0$ hold. Then (6.5) has a unique solution X_S. In addition $N_2(X_S) \le \gamma_S(A,B)N_2(C) \le \hat{\gamma}_S(A,B)N_2(C)$.*

An important example of (6.5) is the Lyapunov equation

$$XA + A^*X = C \qquad (6.8)$$

with $n \times n$ matrices A and C. In this case due to (6.6), with the notation $\theta(A) := \min_k |Re\,\lambda_k(A)|$ we get

$$\|K^{-1}\| \le \hat{\gamma}_L(A) := \sum_{j=0}^{2n-2} \xi_j \frac{g^j(A)}{\theta^{j+1}(A)},$$

where

$$\xi_j := \frac{1}{2^j} \sum_{k=0}^{j} \frac{j!}{((j-k)!k!)^{3/2}}. \qquad (6.9)$$

We thus arrive at.

Corollary 5.9. *Let $\theta(A) > 0$. Then (6.8) has a unique solution X_L and $N_2(X_L) \le \hat{\gamma}_L(A)N_2(C)$. If, in addition, A is normal, then $N_2(X_L) \le N_2(C)/\theta(A)$.*

Corollary 5.7 together with Theorems 2.10 and Corollary 2.6 enables us to investigate perturbations of two-sided matrix Sylvester equations. Moreover, Corollary 5.7 and Theorem 2.11 give us bounds for the derivatives of matrix equations.

5.7   Perturbations of invariant subspaces of finite matrices

Let $A \in \mathbf{C}^{n \times n}$ and $B \in \mathbf{C}^{n \times n}$. Recall that a subspace $S \subset \mathbf{C}^n$ is invariant for A if $AS \subseteq S$. Let the spectra of A and B be separated into two disjoint parts:

$$\sigma(A) = \sigma_1(A) \cup \sigma_2(A) \text{ and } \sigma(B) = \sigma_1(B) \cup \sigma_2(B)$$

with $\sigma_1(A) \cap \sigma_2(A) = \emptyset$, $\sigma_1(B) \cap \sigma_2(B) = \emptyset$, and

$$\delta := \text{dist}(\sigma_2(A), \sigma_1(B)) > 0. \qquad (7.1)$$

Denote by Q_A the orthogonal projection onto the invariant subspace of A, corresponding to $\sigma_1(A)$, and by Q_B the orthogonal projection onto the invariant subspace of B, corresponding to $\sigma_1(B)$. That is, $AQ_A = Q_A AQ_A$, $BQ_B = Q_B AQ_B$, $\sigma(AQ_A) = \sigma_1(A)$ and $\sigma(BQ_B) = \sigma_1(B)$, and put $\hat{Q}_A = I - Q_A$.

A good measure of separation between the subspaces $Q_A \mathbf{C}^n$ and $Q_B \mathbf{C}^n$ is $\|\hat{Q}_A Q_B\|$, cf. [5, p. 202]. In the spectral norm, one has $\|Q_A - Q_B\| = \|\hat{Q}_A Q_B\|$, provided rank $Q_A =$ rank Q_B (see [5, Exercise VII.1.11]).

Put $n_1 =$ rank Q_B and $n_2 =$ rank $\hat{Q}_A$.

Theorem 5.5. *Let condition (7.1) hold. Then*

$$N_2(\hat{Q}_A Q_B) \le N_2(A - B) \sum_{j=0}^{n_1+n_2-2} \frac{j!}{\delta^{j+1}} \sum_{k=0}^{j} \frac{g^k(A)g^{j-k}(B)}{((j-k)!k!)^{3/2}}. \tag{7.2}$$

To prove Theorem 5.5 we need

Lemma 5.8. *Let Q be an orthogonal projection onto an invariant subspace of A. Then $g(AQ) \le g(A)$ and $g((I - Q)A) \le g(A)$.*

Proof. Recall that $A = D_A + V_A$, where D_A is the diagonal part and V_A is the nilpotent part of A. We have $AQ = D_A Q + V_A Q$. It is not hard to check $D_A Q$ is normal and $V_A Q$ is nilpotent, since A, D_A and V_A have the same invariant subspaces. Consequently, $g(AQ) = N_2(V_A Q) \le N_2(V_A) = g(A)$. Similarly the second inequality can be proved. Q. E. D.

Proof of Theorem 5.5: Note that

$$\hat{Q}_A A = \hat{Q}_A A(Q_A + \hat{Q}_A) = \hat{Q}_A Q_A AQ_A + \hat{Q}_A A\hat{Q}_A = \hat{Q}_A A\hat{Q}_A$$

and denote $A_2 = \hat{Q}_A A, B_1 = BQ_B, C_{21} = \hat{Q}_A(A - B)Q_B$ and $X_{21} = \hat{Q}_A Q_B$. Then

$$C_{21} = \hat{Q}_A A\hat{Q}_A Q_B - \hat{Q}_A Q_B BQ_B = \hat{Q}_A AX_{21} - X_{21}BQ_B,$$

and therefore,

$$A_2 X_{21} - X_{21}B_1 = C_{21}. \tag{7.3}$$

Clearly,

$$N_2(C_{21}) = N_2(\hat{Q}_A(A - B)Q_B) \le N_2(A - B)\|\hat{Q}_A\|\|Q_B\| = N_2(A - B). \tag{7.4}$$

Apply to (7.3) Corollary 5.8 with $A = A_2, B = B_1$. Taking into account that $\rho_S(A_2, B_1) = \delta$, we have

$$N_2(X_{21}) \leq N_2(C_{21}) \sum_{j=0}^{n_1+n_2-2} \frac{j!}{\delta^{j+1}} \sum_{k=0}^{j} \frac{g^k(A_2)g^{j-k}(B_1)}{((j-k)!k!)^{3/2}}.$$

Due to the previous lemma, $g(A_2) \leq g(A), g(B_1) \leq g(B)$. This, (7.4) and the equality $X_{21} = \hat{Q}_A Q_B$ prove the theorem. Q. E. D.

If A is normal, then (7.2) yields the inequality

$$N_2(\hat{Q}_A Q_B) \leq N_2(A - B) \sum_{j=0}^{n_2-1} \frac{g^j(B)}{\sqrt{j!}\delta^{j+1}}. \tag{7.5}$$

If both A and B are normal, then (7.2) yields

$$N_2(\hat{Q}_A Q_B) \leq \frac{N_2(A - B)}{\delta}.$$

This is the result of [15] (in the finite dimensional case).

5.8 Comments to Chapter 5

The results in Sections 5.1-5.6 is taken from the paper [48]. Theorem 5.5 is adopted from [49].

Chapter 6

Bounds for Condition Numbers of Diagonalizable Matrices

An eigenvalue is said to be simple, if its geometric multiplicity is equal to one. In this chapter we consider a matrix A whose all the eigenvalues are simple. As it is well known, in this case there is an invertible matrix T, such that

$$T^{-1}AT = \hat{D}, \qquad (0.1)$$

where $\hat{D}$ is a normal matrix. Besides, A is called a diagonalizable matrix. The condition number $\kappa(A, T) := \|T\|\|T^{-1}\|$ is very important for various applications. We obtain a bound for the condition number and discuss applications of that bound to matrix functions and equations with diagonalizable matrices.

6.1 A bound for condition numbers of matrices

If $A \in \mathbf{C}^{n \times n}$ $(n \geq 2)$ is diagonalizable, it can be written as

$$A = \sum_{k=1}^{n} \lambda_k \hat{Q}_k \in \mathbf{C}^{n \times n} \ (\lambda_k = \lambda_k(A) \in \sigma(A)),$$

where $\hat{Q}_k$ are one-dimensional eigen-projections. If $f(z)$ is a scalar function defined on the spectrum of A, then $f(A)$ is defined as

$$f(A) = \sum_{k=1}^{n} f(\lambda_k)\hat{Q}_k.$$

Let

$$r(z) = \sum_{k=0}^{n} c_k z^k \ (z \in \mathbf{C})$$

105

be a polynomial, such that $r(\lambda_k) = f(\lambda_k)$ $(\lambda_k = \lambda_k(A);\ k = 1,...,n)$. Then

$$f(A) = r(A) = \sum_{k=0}^{n} c_k A^k$$

(see [20, Section V.1]). From (0.1) it follows

$$f(A) = \sum_{k=0}^{n} c_k A^k = \sum_{k=0}^{n} c_k T^{-1} \hat{D}^k T = T^{-1} f(\hat{D}) T.$$

Since $\hat{D}$ is normal, $\|f(\hat{D})\| = \max_k |f(\lambda_k)|$. We thus arrive at

Lemma 6.1. *Let condition (0.1) hold and $f(z)$ be a scalar function defined on the $\sigma(A)$ for an $A \in \mathbf{C}^{n \times n}$. Then*

$$\|f(A)\| \le \kappa(A, T) \max_k |f(\lambda_k)|.$$

In particular,

$$\|A^m\| \le \kappa(A, T) r_s^m(A) \quad (m = 1, 2, ...),$$

$$\|e^{At}\| \le \kappa(A, T) e^{\alpha(A)t} \quad (t \ge 0),$$

$$\|(A - \lambda I)^{-1}\| \le \frac{\kappa(A, T)}{\rho(A, \lambda)} \quad (\lambda \notin \sigma(A))$$

and

$$\|A^{-m}\| \le \kappa(A, T) \frac{1}{r_{\text{low}}^m(A)} \quad (m = 1, 2, ...).$$

We are going to estimate the condition number of A under the condition

$$\lambda_j \ne \lambda_m \text{ whenever } j \ne m. \tag{1.1}$$

Recall that

$$g(A) := (N_2^2(A) - \sum_{k=1}^{n} |\lambda_k|^2)^{1/2}$$

(see Section 3.1). Put

$$\delta_j := \min_{k=1,...,n;\ k \ne j} |\lambda_j - \lambda_k|, \tau_j(A) := \sum_{k=0}^{n-2} \frac{g^k(A)}{\sqrt{k!} \delta_j^{k+1}}$$

and

$$\gamma(A) := \left(1 + \frac{g(A)}{n-1} \sqrt{\sum_{j=1}^{n-1} \tau_j^2(A)}\right)^{2(n-1)}.$$

Theorem 6.1. *Let condition (1.1) be fulfilled. Then there is an invertible matrix T, such that (0.1) holds with*

$$\kappa(A, T) \le \gamma(A). \tag{1.2}$$

The proof of this theorem is presented in the next two sections. Theorem 6.1 is sharp: if A is normal, then $g(A) = 0$ and $\gamma(A) = 1$. Thus we obtain the equality $\kappa_T = 1$.

Lemma 6.1 and Theorem 6.1 immediately imply.

Corollary 6.1. *Let condition (1.1) hold and $f(z)$ be a scalar function defined on the spectrum of A. Then $\|f(A)\| \leq \gamma(A) \max_k |f(\lambda_k)|$.*

6.2 Auxiliary results

Let matrix A have in $\mathbf{C}^n$ a chain of invariant projections P_k $(k = 1, ..., m;$ $m \leq n)$:

$$0 \subset P_1\mathbf{C}^n \subset P_2\mathbf{C}^n \subset ... \subset P_m\mathbf{C}^n = \mathbf{C}^n \tag{2.1}$$

and

$$P_k A P_k = A P_k \quad (k = 1, ..., m). \tag{2.2}$$

Put $\Delta P_k = P_k - P_{k-1}$ $(P_0 = 0)$, $Q_k = I - P_k$; denote by A_k the restriction of $\Delta P_k A \Delta P_k$ onto $\Delta P_k \mathbf{C}^n$ $(k = 1, ..., m)$, and by B_j the restriction of $Q_k A Q_k$ onto $Q_j \mathbf{C}^n$ $(j = 1, ..., m - 1)$.

Lemma 6.2. *One has*

$$\sigma(A) = \cup_{k=1}^m \sigma(A_k).$$

Proof. Put

$$S = \sum_{k=1}^m A_k \text{ and } W = A - S.$$

Due to (2.2) we have $W P_k = P_{k-1} W P_k$. Hence,

$$W^m = W^m P_m = W^{m-1} P_{m-1} W P_m = W^{m-2} P_{m-2} W P_{m-1} W P_m =$$

$$W^{m-2} P_{m-2} W^2 = W^{m-3} P_{m-3} W^3 = ... = P_0 W^m = 0.$$

So W is nilpotent. Similarly, taking into account that

$$(S - \lambda I)^{-1} W P_k = P_{k-1}(S - \lambda I)^{-1} W P_k$$

we prove that $((S - \lambda I)^{-1} W)^m = 0$ $(\lambda \notin \sigma(S))$. Thus

$$(A - \lambda I)^{-1} = (S + W - \lambda I)^{-1} = (I + (S - \lambda I)^{-1} W)^{-1}(S - \lambda I)^{-1} =$$

$$\sum_{k=0}^{m-1} (-1)^k ((S - \lambda I)^{-1} W)^k (S - \lambda I)^{-1}.$$

Hence it easily follows that $\sigma(S) = \sigma(A)$. This proves the lemma. Q. E. D.

Let the spectra $\sigma(A_k)$ of A_k satisfy the condition

$$\sigma(A_k) \cap \sigma(A_j) = \emptyset \quad (j \neq k; \; j, k = 1, ..., m). \tag{2.3}$$

According to the previous lemma we have

$$\sigma(B_j) = \cup_{k=j+1}^{m} \sigma(A_k) \quad (j = 1, ..., m-1).$$

So due to (2.3)

$$\sigma(B_j) \cap \sigma(A_j) = \emptyset.$$

Under this condition, making use of Theorem 2.2 (see also Section 2.3), we can assert that the equation

$$A_j X_j - X_j B_j = -C_j \quad (j = 1, ..., m-1), \tag{2.4}$$

where $C_j = \Delta P_j A Q_j$, has a unique solution.

Lemma 6.3. *Let condition (2.3) hold and X_j be a solution to (2.4). Then*

$$(I - X_{m-1})(I - X_{m-2}) \cdots (I - X_1) A (I + X_1)(I + X_2) \cdots (I + X_{m-1}) =$$

$$A_1 + A_2 + ... + A_m. \tag{2.5}$$

Proof. Since $X_j = \Delta P_j X_j Q_j$, we have

$$X_j A_j = B_j X_j = X_j C_j = C_j X_j = 0.$$

Due to (2.2), $Q_j A P_j = 0$. Thus, $A = A_1 + B_1 + C_1$ and consequently,

$$(I - X_1)A(I + X_1) = (I - X_1)(A_1 + B_1 + C_1)(I + X_1) =$$

$$A_1 + B_1 + C_1 - X_1 B_1 + A_1 X_1 = A_1 + B_1.$$

Furthermore, $B_1 = A_2 + B_2 + C_2$. Hence,

$$(Q_1 - X_2)B_1(Q_1 + X_2) = (Q_1 - X_1)(A_2 + B_2 + C_2)(Q_1 + X_1) =$$

$$A_2 + B_2 + C_2 - X_2 B_2 + A_2 X_2 = A_2 + B_2.$$

Therefore,

$$(I - X_2)(A_1 + B_1)(I + X_2) = (P_1 + Q_1 - X_2)(A_1 + B_1)(P_1 + Q_1 + X_2) =$$

$$A_1 + (Q_1 - X_2)(A_1 + B_1)(Q_1 + X_2) = A_1 + A_2 + B_2.$$

Consequently,

$$(I - X_2)(A_1 + B_1)(I + X_2) = (I - X_2)(I - X_1)A(I + X_1)(I + X_2)$$

$$= A_1 + A_2 + B_2.$$

Continuing this process and taking into account that $B_{m-1} = A_m$, we obtain the required result. Q. E. D.

Take

$$T = (I + X_1)(I + X_2) \cdots (I + X_{m-1}). \tag{2.6}$$

It is simple to see that the inverse to $I + X_j$ is the matrix $I - X_j$. Thus,

$$T^{-1} = (I - X_{m-1})(I - X_{m-2}) \cdots (I - X_1) \tag{2.7}$$

and (2.5) can be written as

$$T^{-1}AT = diag\, (A_k)_{k=1}^m. \tag{2.8}$$

By the inequalities between the arithmetic and geometric means we get

$$\|T\| \leq \prod_{k=1}^{m-1} (1 + \|X_k\|) \leq \left(1 + \frac{1}{m-1} \sum_{k=1}^{m-1} \|X_k\|\right)^{m-1} \tag{2.9}$$

and

$$\|T^{-1}\| \leq \left(1 + \frac{1}{m-1} \sum_{k=1}^{m-1} \|X_k\|\right)^{m-1}. \tag{2.10}$$

6.3   Proof of Theorem 6.1

Let $\{e_k\}$ be the Schur basis (the orthogonal normal basis of the triangular representation) of matrix A:

$$A = \begin{pmatrix} a_{11} & a_{12} & a_{13} & \cdots & a_{1n} \\ 0 & a_{22} & a_{23} & \cdots & a_{2n} \\ \cdot & \cdot & \cdot & \cdots & \cdot \\ 0 & 0 & 0 & \cdots & a_{nn} \end{pmatrix}$$

with $a_{jj} = \lambda_j$ in that basis. Besides,

$$\sum_{k=2}^{n} \sum_{i=1}^{k-1} |a_{ik}|^2 = g^2(A).$$

Take $P_j = \sum_{k=1}^j (., e_k) e_k$. Then one can apply Lemma 6.3 with $m = n$, $\Delta P_k = (., e_k) e_k$,

$$Q_j = \sum_{k=j+1}^n (., e_k) e_k, A_k = \lambda_k \Delta P_k, diag\,(A_k)_{k=1}^n =$$

$$diag\,(\lambda_k)_{k=1}^n,$$

$$B_j = Q_j A Q_j = \begin{pmatrix} a_{j+1,j+1} & a_{j+1,j+2} & \cdots & a_{j+1,n} \\ 0 & a_{j+2,j+2} & \cdots & a_{j+2,n} \\ . & . & \cdots & . \\ 0 & 0 & . & a_{nn} \end{pmatrix}$$

and

$$C_j = \Delta P_j A Q_j = \begin{pmatrix} a_{j,j+1} & a_{j,j+2} & \cdots & a_{j,n} \end{pmatrix}.$$

Besides,

$$A = \begin{pmatrix} \lambda_1 & C_1 \\ 0 & B_1 \end{pmatrix}, B_1 = \begin{pmatrix} \lambda_2 & C_2 \\ 0 & B_2 \end{pmatrix}, ..., B_j = \begin{pmatrix} \lambda_{j+1} & C_{j+1} \\ 0 & B_{j+1} \end{pmatrix} \quad (j = 1, ..., n-1).$$

So B_j is an upper-triangular $(n - j) \times (n - j)$-matrix. Equation (2.4) takes the form $\lambda_j X_j - X_j B_j = -C_j$. Since $X_j = X_j Q_j$, we can write $X_j(\lambda_j Q_j - B_j) = C_j$. Therefore

$$X_j = C_j (\lambda_j Q_j - B_j)^{-1}. \tag{3.1}$$

The inverse matrix is understood in the sense of subspace $Q_j \mathbf{C}^n$. Hence,

$$\|X_j\| \le \|C_j\| \|(\lambda_j Q_j - B_j)^{-1}\|.$$

Besides,

$$\|C_j\|^2 = \sum_{k=j+1}^n |a_{jk}|^2$$

and due to Theorem 3.2, and the inequality $\rho(B_j, \lambda_j) \ge \delta_j$, we have

$$\|(\lambda_j Q_j - B_j)^{-1}\| \le \sum_{k=0}^{n-j-1} \frac{g^k(B_j)}{\sqrt{k!}\delta_j^{k+1}} \quad (j = 1, 2, ..., n-1).$$

But

$$g^2(B_j) = g^2(Q_j A Q_j) = \sum_{k=j+2}^n \sum_{i=j+1}^{k-1} |a_{ik}|^2 \le g^2(A).$$

So

$$\|(\lambda_j Q_j - B_j)^{-1}\| \le \tau_j(A) \text{ and } \|X_j\| \le \|C_j\|\tau_j(A).$$

Take T as is in (2.6) with X_k defined by (3.1). Besides (2.9) and (2.10) imply

$$\|T\| \le \left(1 + \frac{1}{n-1}\sum_{k=1}^{n-1}\|X_k\|\right)^{n-1} \le \left(1 + \frac{1}{n-1}\sum_{k=1}^{n-1}\tau_k(A)\|C_k\|\right)^{n-1}.$$

and

$$\|T^{-1}\| \le \left(1 + \frac{1}{n-1}\sum_{j=1}^{n-1}\tau_j(A)\|C_j\|\right)^{n-1}.$$

But by the Schwarz inequality,

$$(\sum_{j=1}^{n-1}\tau_j(A)\|C_j\|)^2 \le \sum_{k=1}^{n-1}\tau_k^2(A)\sum_{j=1}^{n-1}\|C_j\|^2 = \sum_{k=1}^{n-1}\tau_k^2(A)\sum_{j=1}^{n-1}\sum_{i=j+1}^{n}|a_{ji}|^2 =$$

$$\sum_{k=1}^{n-1}\tau_k^2(A)g^2(A).$$

Thus

$$\|T\|^2 \le \left(1 + \frac{g(A)}{n-1}\sqrt{\sum_{k=1}^{n-1}\tau_k^2(A)}\right)^{2(n-1)} = \gamma(A)$$

and $\|T^{-1}\|^2 \le \gamma(A)$. Now (2.8) proves the theorem. Q. E. D.

6.4 The quasi-Sylvester equation with diagonalizable matrices

Consider the equation

$$Y - AYB = C \qquad (4.1)$$

with $A \in \mathbf{C}^{n\times n}$, $B \in \mathbf{C}^{\tilde{n}\times\tilde{n}}$ and $C \in \mathbf{C}^{n\times\tilde{n}}$. It is assumed that A and B are diagonalizable matrices. So, in addition to (0.1) there is an invertible matrix $\tilde{T}$ and a normal matrix $\tilde{D}$, such that

$$\tilde{T}^{-1}B\tilde{T} = \tilde{D}.$$

Putting $\kappa(B, \tilde{T}) = \|\tilde{T}\| \|\tilde{T}^{-1}\|$, we have
$$\|A^m\| \le \kappa(A, T)r_s^m(A), \|B^m\| \le \kappa(B, \tilde{T})r_s^m(B) \quad (m = 1, 2, ...). \quad (4.2)$$
Under the condition
$$r_s(A)r_s(B) < 1 \quad (4.3)$$
as is shown in Section 2.2, equation (4.1) has a unique solution Y which can be represented as
$$Y = \sum_{k=0}^{\infty} A^k C B^k.$$
Now (4.2) implies
$$\|Y\| \le \|C\| \sum_{t=0}^{\infty} \kappa(A, T)\kappa(B, \tilde{T})r_s^t(A)r_s^t(B).$$
Hence we get

Corollary 6.2. *Let both matrices A and B be diagonalizable and condition (4.3) hold. Then the unique solution Y of (4.1) is subject to the inequality*
$$\|Y\| \le \frac{\kappa(A, T)\kappa(B, \tilde{T})\|C\|}{1 - r_s(A)r_s(B)} \quad (4.4)$$

Let $\tilde{\lambda}_k \ (k = 1, ..., \tilde{n})$ be the eigenvalues of B and
$$\tilde{\lambda}_j \ne \tilde{\lambda}_m \text{ whenever } j \ne m. \quad (4.5)$$
Then one can apply Theorem 6.1 to estimate $\kappa(B, \tilde{T})$.

6.5   The Sylvester equation with diagonalizable matrices

Consider the equation
$$AX - XB = C, \quad (5.1)$$
with $A \in \mathbf{C}^{n \times n}$, $B \in \mathbf{C}^{\tilde{n} \times \tilde{n}}$ and $C \in \mathbf{C}^{n \times \tilde{n}}$.

Assume that
$$\beta(B) > \alpha(A). \quad (5.2)$$
Then due to Theorem 2.4
$$X = -\int_0^{\infty} e^{At} C e^{-Bt} dt$$
If both matrices A and B are diagonalizable, then be obtain
$$\|X\| \le \kappa(A, T)\kappa(B, \tilde{T})\|C\| \int_0^{\infty} e^{-\beta(B)t} e^{\alpha(A)t} dt.$$
Hence we get

Corollary 6.3. *Let both matrices A and B be diagonalizable and condition (5.2) hold. Then the unique solution X of (5.1) is subject to the inequality*
$$\|X\| \le \frac{\kappa(A, T)\kappa(B, \tilde{T})\|C\|}{\beta(B) - \alpha(A)}. \quad (5.3)$$

6.6 Comments to Chapter 6

The material in this chapter is based on the paper [38].

Chapter 7

Functions of a Compact Operator in a Hilbert Space

This chapter is devoted to functions of some classes of compact operators in a Hilbert space. In particular, we establish norm estimates for the resolvent of a Schatten-von Neumann operator and for functions of a Hilbert-Schmidt operator, regular on the convex hull of the spectrum. Applications to spectrum perturbations of compact operators and to operator equations whose coefficients are compact operators are also discussed.

7.1 Schatten-von Neumann operators

Recall that $\mathcal{H}$ is a separable Hilbert space with a scalar product $(.,.)$, the norm $\|.\|_{\mathcal{H}} = \|.\|$ and unit operator I; $\mathcal{B}(\mathcal{H})$ is the algebra of all bounded linear operators in $\mathcal{H}$. For a compact operator $A \in \mathcal{B}(\mathcal{H})$, $\lambda_k(A)$ $(k = 1, 2, ...)$ are the eigenvalues of A, taken with their multiplicities and ordered in the decreasing way: $|\lambda_k(A)| \geq |\lambda_{k+1}(A)|$ and $s_k(A)$ $(k = 1, 2, ...)$ are the singular numbers of A, taken with their multiplicities and ordered in the decreasing way. Recall that the singular numbers of A are the eigenvalues of $(A^*A)^{1/2}$. Throughout this section A and B are compact operators in $\mathcal{H}$.

Lemma 7.1. *If C and D are bounded linear operators in $\mathcal{H}$, then*
$$s_k(CAD) \leq \|C\|\|D\|s_k(A) \quad (k \geq 1).$$

For the proof see Proposition IV.2.3 from [53].

The following results are also well known, cf. [54, Section II.4.2], [53, Section IV.4].

Lemma 7.2. *One has*
$$\sum_{k=1}^{j} s_k(A + B) \leq \sum_{k=1}^{j} (s_k(A) + s_k(B)),$$

$$\sum_{k=1}^{j} s_k^p(AB) \le \sum_{k=1}^{j} s_k^p(A) s_k^p(B) \quad (p > 0)$$

and

$$\prod_{k=1}^{j} s_k(AB) \le \prod_{k=1}^{j} s_k(A) s_k(B) \quad (j = 1, 2, \ldots).$$

Recall that A is said to be normal if $AA^* = A^*A$.

Lemma 7.3. *(Weyl's inequalities) The inequalities*

$$\prod_{j=1}^{k} |\lambda_j(A)| \le \prod_{j=1}^{k} s_j(A)$$

and

$$\sum_{j=1}^{k} |\lambda_j(A)| \le \sum_{j=1}^{k} s_j(A) \quad (k = 1, 2, \ldots)$$

are true. They become equalities if and only if A is normal.

For the proof see Theorem IV.3.1 and Corollary IV.3.4 from [53], and Section II.3.1 from [54]).

The set of compact operators $A \in \mathcal{B}(\mathcal{H})$ satisfying the condition

$$N_p(A) = \left[\sum_{k=1}^{\infty} s_k^p(A) \right]^{1/p} < \infty$$

for some $p \in [1, \infty)$, is called *the Schatten-von Neumann ideal and is denoted by SN_p.* $N_p(.)$ is called the Schatten-von Neumann p-norm. Besides, SN_1 is the ideal of *nuclear operators (the Trace class)* and SN_2 is the ideal of *Hilbert-Schmidt operators*, $N_2(A)$ is the Hilbert-Schmidt norm.

From Lemma 7.1 we have

$$N_p(DAC) \le N_p(A) \|D\| \|C\| \quad (A \in SN_p; C, D \in \mathcal{B}(\mathcal{H})). \tag{1.1}$$

The following propositions are true (the proofs can be found for instance, in the books [54, Section III.7], [17]).

Lemma 7.4. *If $A \in SN_p$ and $B \in SN_q$ $(1 < p, q < \infty)$, then $AB \in SN_s$ with $1/s = 1/p + 1/q$. Moreover, $N_s(AB) \le N_p(A)N_q(B)$.*

Let $\{e_k\}$ be an orthogonal normal basis in $\mathcal{H}$ and the series

$$\sum_{k=1}^{\infty} (Ae_k, e_k)$$

converges. Then the sum of this series is called *the trace of A*:

$$\text{trace } A = \sum_{k=1}^{\infty} (Ae_k, e_k).$$

It is well known that

$$N_p(A) = \sqrt[p]{\text{trace } (AA^*)^{p/2}}.$$

The Schatten-von Neumann p-norms are non-increasing in p. In other words, for $1 \leq p \leq s$ we have $N_p(A) \geq N_s(A)$, provided $A \in SN_p$.

The Schatten-von Neumann norm is unitarily invariant. This means that $N_p(UAU_1) = N_p(A)$ for any choice of linear unitary operators U and U_1.

For all $p, q \in (1, \infty)$, satisfying the equation $\frac{1}{p} + \frac{1}{q} = 1$ we have

$$N_p(A) = \sup \{\text{trace } (AB) : B \in SN_q, N_q(B) \leq 1\}$$

and

$$|\text{trace } (AB)| \leq N_p(A)N_q(B).$$

From the Weyl inequalities it directly follows that

$$\sum_{j=1}^{\infty} |\lambda_j(A)|^p \leq N_p^p(A) \quad (A \in SN_p, 1 \leq p < \infty). \tag{1.2}$$

Hence, due Lemma 1.8 we arrive at the following result.

Corollary 7.1. *Let $A_n, A \in SN_p$ $(1 \leq p < \infty : n = 1, 2, ...)$, and let $A_n \to A$ in the norm $N_p(.)$. Then*

$$\sum_{k=1}^{\infty} |\lambda_k(A_n)|^p \to \sum_{k=1}^{\infty} |\lambda_k(A)|^p, \text{ as } n \to \infty$$

7.2 The resolvent of a Hilbert-Schmidt operator

Let A be a Hilbert-Schmidt operator. The following quantity plays a key role in this section:

$$g(A) = [N_2^2(A) - \sum_{k=1}^{\infty} |\lambda_k(A)|^2]^{1/2}.$$

Recall that for a finite dimensional operator A, $g(A)$ is defined in Section 3.1. Since

$$\sum_{k=1}^{\infty} |\lambda_k(A)|^2 \geq |\sum_{k=1}^{\infty} \lambda_k^2(A)| = |Trace\ A^2|,$$

one can write

$$g^2(A) \leq N_2^2(A) - |Trace\ A^2|. \tag{2.1}$$

If A is a normal Hilbert-Schmidt operator, then $g(A) = 0$, since

$$N_2^2(A) = \sum_{k=1}^{\infty} |\lambda_k(A)|^2$$

in this case. Furthermore, Corollary 7.1 implies

Corollary 7.2. *Let $A, A_n \in SN_2$ ($n = 1, 2, ..$) and $A_n \to A$ in the norm $N_2(.)$. Then*

$$\lim_{n \to \infty} g(A_n) = g(A).$$

Let $A_I = \Im A = (A - A^*)/2i$. Due to the previous corollary and Theorem 3.1 we can write

$$g^2(A) = 2N_2^2(A_I) - 2\sum_{k=1}^{\infty} (\Im \lambda_k(A))^2 \leq 2N_2^2(A_I) \tag{2.2}$$

for any $A \in SN_2$.

Recall that $\rho(A, \lambda) := \inf_{t \in \sigma(A)} |\lambda - t|$.

Theorem 7.1. *Let A be a Hilbert-Schmidt operator. Then the inequalities*

$$\|R_\lambda(A)\| \leq \sum_{k=0}^{\infty} \frac{g^k(A)}{\rho^{k+1}(A, \lambda)\sqrt{k!}} \tag{2.3}$$

and

$$\|R_\lambda(A)\| \leq \frac{1}{\rho(A, \lambda)} exp \left[\frac{1}{2} + \frac{g^2(A)}{2\rho^2(A, \lambda)} \right] \quad (\lambda \notin \sigma(A)) \tag{2.4}$$

are true.

Proof. Let A_n ($n = 1, 2, ...$) be a sequence of n-dimensional operators converging to A in the norm N_2. Then by Theorem 3.2,

$$\|R_\lambda(A_n)\| \leq \sum_{k=0}^{n-1} \frac{g^k(A_n)}{\rho^{k+1}(A_n, \lambda)\sqrt{k!}} \quad (\lambda \notin \sigma(A_n)).$$

Now Corollary 7.2 and Lemma 1.6 prove inequality (2.3).

Inequality (2.4) is due to Theorem 3.4 and Lemma 1.6, since

$$(1 + \frac{x}{n})^n \leq e^x \quad (x \geq 0; \ n = 1, 2, ...).$$

Q. E. D.

Theorem 7.1. is sharp: if A is a normal operator, then $g(A) = 0$ and from (2.3) we have $\|R_\lambda(A)\| = \frac{1}{\rho(A,\lambda)}$.

7.3 Resolvents of Schatten-von Neumann operators

Theorem 7.2. *For some integer $p \geq 2$, let*

$$A \in SN_{2p}. \tag{3.1}$$

Then

$$\|R_\lambda(A)\| \leq \sum_{m=0}^{p-1} \sum_{k=0}^{\infty} \frac{(2N_{2p}(A))^{pk+m}}{\rho^{pk+m+1}(A,\lambda)\sqrt{k!}} \quad (\lambda \notin \sigma(A)). \tag{3.2}$$

The proof of this theorem is presented in the next section. In the next section we also prove the following result.

Theorem 7.3. *Under condition (3.1) one has*

$$\|R_\lambda(A)\| \leq \sqrt{e} \sum_{m=0}^{p-1} \frac{(2N_{2p}(A))^m}{\rho^{m+1}(A,\lambda)} \exp\left[\frac{(2N_{2p}(A))^{2p}}{2\rho^{2p}(A,\lambda)}\right] \quad (\lambda \notin \sigma(A)). \tag{3.3}$$

Remark 7.1. Since, condition (3.1) implies $A - A^* \in SN_{2p}$, additional estimates for the resolvent established in Chapter 9 below can be applied.

Note that if (3.1) holds, then

$$A^p \text{ is a Hilbert-Schmidt operator.} \tag{3.4}$$

Use the identity

$$A^p - I\lambda^p = (A - I\lambda)\sum_{k=0}^{p-1} A^k \lambda^{p-k-1} = (A - I\lambda)T_{\lambda,p} \quad (\lambda^p \notin \sigma(A^p)),$$

where

$$T_{\lambda,p} = \sum_{k=0}^{p-1} A^k \lambda^{p-k-1}.$$

This implies

$$(A - I\lambda)^{-1} = T_{\lambda,p}(A^p - I\lambda^p)^{-1}. \tag{3.5}$$

Thus,

$$\|(A - I\lambda)^{-1}\| \le \|T_{\lambda,p}\| \, \|(A^p - I\lambda^p)^{-1}\|.$$

Applying inequality (2.3) to the expression $(A^p - I\lambda^p)^{-1} = R_{\lambda^p}(A^p)$, we obtain:

$$\|R_{\lambda^p}(A^p)\| \le \sum_{k=0}^{\infty} \frac{g^k(A^p)}{\rho^{k+1}(A^p, \lambda^p)\sqrt{k!}} \quad (\lambda^p \notin \sigma(A^p)),$$

where

$$\rho(A^p, \lambda^p) = \inf_{t \in \sigma(A)} |t^p - \lambda^p|$$

is the distance between $\sigma(A^p)$ and the point λ^p. This implies

$$\|R_{\lambda}(A)\| \le \|T_{\lambda,p}\| \sum_{k=0}^{\infty} \frac{g^k(A^p)}{\rho^{k+1}(A^p, \lambda^p)\sqrt{k!}} \quad (\lambda^p \notin \sigma(A^p)), \tag{3.6}$$

provided condition (3.4) holds.

Similarly, making use of inequality (2.4), under condition (3.4), we obtain

$$\|R_{\lambda}(A)\| \le \frac{\sqrt{e}\|T_{\lambda,p}\|}{\rho(A^p, \lambda^p)} \, exp \, \Big[\frac{g^2(A^p)}{2\rho^2(A^p, \lambda^p)}\Big] \quad (\lambda^p \notin \sigma(A^p)). \tag{3.7}$$

7.4   Proofs of Theorems 7.2 and 7.3

We need the following result.

Lemma 7.5. *Let A be a linear operator acting in a Euclidean space $\mathbf{C}^n$ for $n = jp$ with integers $p \ge 1, j > 1$. Then*

$$\|R_{\lambda}(A)\| \le \sum_{m=0}^{p-1} \sum_{k=0}^{j} \frac{N_{2p}^{kp+m}(V)}{\rho^{pk+m+1}(A, \lambda)\sqrt{k!}} \quad (\lambda \notin \sigma(A)), \tag{4.1}$$

where V is the nilpotent part of A.

Proof. Since $A = D + V$ $(\sigma(A) = \sigma(D))$, where D is the diagonal part of A,

$$(A - \lambda I)^{-1} = (D + V - \lambda I)^{-1} = (D - \lambda I)^{-1}(I - B_{\lambda})^{-1} \tag{4.2}$$

where $B_{\lambda} = -VR_{\lambda}(D)$. By the identity

$$(I - B_{\lambda})(I + B_{\lambda} + \ldots + B_{\lambda}^{p-1}) = I - B_{\lambda}^p$$

we have

$$(A - \lambda I)^{-1} = (D + V - \lambda I)^{-1} =$$

$$(D - \lambda I)^{-1}(I + B_\lambda + \ldots + B_\lambda^{p-1})(I - B_\lambda^p)^{-1}. \tag{4.3}$$

Clearly, B_λ is a nilpotent operator. So $B_\lambda^n = B_\lambda^{pj} = 0$ and one can write

$$(I - B_\lambda^p)^{-1} = \sum_{k=0}^{j} B_\lambda^{kp}.$$

Thus,

$$(I - B_\lambda)^{-1} = \sum_{m=0}^{p-1} \sum_{k=0}^{j} B_\lambda^{kp+m} \ .$$

Hence,

$$R_\lambda(A) = R_\lambda(D) \sum_{m=0}^{p-1} \sum_{k=0}^{j} B_\lambda^{kp+m}. \tag{4.4}$$

Take into account that

$$N_2(B_\lambda^p) \le N_{2p}^p(B_\lambda) \ \ (\lambda \notin \sigma(A)). \tag{4.5}$$

Lemma 3.4 gives us the inequality

$$\|B_\lambda^{pk}\| \le \frac{N_2^k(B_\lambda^p)}{\sqrt{k!}}.$$

Therefore, (4.5) yields the estimate

$$\|B_\lambda^{pk}\| \le \frac{N_{2p}^{kp}(B_\lambda)}{\sqrt{k!}}.$$

Furthermore, it is clear that

$$N_{2p}(B_\lambda) = N_{2p}(V R_\lambda(D)) \le N_{2p}(V)\|R_\lambda(D)\|.$$

Since D is normal and $\sigma(A) = \sigma(D)$,

$$\|R_\lambda(D)\| = \rho^{-1}(D, \lambda) = \rho^{-1}(A, \lambda). \tag{4.6}$$

Hence,

$$N_{2p}(B_\lambda) \le \frac{N_{2p}(V)}{\rho(A, \lambda)}. \tag{4.7}$$

Thus,

$$\|B_\lambda^{pk}\| \le \frac{N_{2p}^{kp}(V)}{\rho^{kp}(A, \lambda)\sqrt{k!}}.$$

Evidently, $\|B_\lambda^m\| \leq N_{2p}^m(B_\lambda)$. Now relation (4.7) implies

$$\|B_\lambda^m\| \leq \frac{N_{2p}^m(V)}{\rho^m(A,\lambda)}.$$

Consequently,

$$\|B_\lambda^{pk+m}\| \leq \frac{N_{2p}^{kp+m}(V)}{\rho^{kp+m}(A,\lambda)\sqrt{k!}}.$$

Taking into account (4.4), we arrive at the inequality

$$\|R_\lambda(A)\| \leq \|R_\lambda(D)\| \sum_{m=0}^{p-1} \sum_{k=0}^{j} \frac{N_{2p}^{kp+m}(V)}{\rho^{kp+m}(A,\lambda)\sqrt{k!}}.$$

Now (4.6) yields the required result. Q. E. D.

Corollary 7.3. *Let $A \in \mathbf{C}^{n \times n}$ $(p = 1, 2, ...)$. Then inequality (3.2) is valid.*

Indeed, inequality (1.2) implies $N_{2p}(D) \leq N_{2p}(A)$. Thus, the triangular representation yields

$$N_{2p}(V) \leq N_{2p}(A) + N_{2p}(D) \leq 2N_{2p}(A). \tag{4.8}$$

Now required result is due the previous lemma.

Proof of Theorem 7.2: Making use of Corollary 7.3 and Lemma 1.6, we prove (3.2), since any operator from the Schatten-von Neumann ideal is a limit of finite dimensional operators in the norm of that ideal. Q. E. D.

To prove Theorem 7.3, we need the following

Lemma 7.6. *Let V be the nilpotent part of $A \in \mathbf{C}^{n \times n}$. Then for any integer $p \geq 1$,*

$$\|R_\lambda(A)\| \leq \sqrt{e} \sum_{m=0}^{p-1} \frac{(N_{2p}(V))^m}{\rho^{m+1}(A,\lambda)} \, exp\left[\frac{(N_{2p}(V))^{2p}}{2\rho^{2p}(A,\lambda)}\right] \; (\lambda \notin \sigma(A)).$$

Proof. According to (4.3),

$$\|(A - \lambda I)^{-1}\| \leq \|(D - \lambda I)^{-1}\| \sum_{k=0}^{p-1} \|B_\lambda\|^k \|(I - B_\lambda^p)^{-1}\|. \tag{4.9}$$

Besides, B_λ^p is nilpotent. Due to (2.4),

$$\|(I - B_\lambda^p)^{-1}\| \leq e^{1/2} e^{N_2^2(B_\lambda^p)/2}.$$

But due to (4.7) we can write

$$N_2(B_\lambda^p) \leq N_{2p}^p(B_\lambda) \leq N_{2p}^p(V)\|R_\lambda(D)\|^p \leq \frac{N_{2p}^p(V)}{\rho^p(A, \lambda)}.$$

Thus,

$$\|(I - B_\lambda^p)^{-1}\| \leq e^{1/2}exp\,[\frac{N_{2p}^{2p}(V)}{2\rho^{2p}(A, \lambda)}].$$

Now (4.9) implies the result. Q. E. D.

Proof of Theorem 7.3: From the preceding lemma and (4.8) we have inequality (3.3) for a finite dimensional operator. To finish the proof we need only to apply Lemma 1.6, since any operator from the Schatten-von Neumann ideal is a limit of finite dimensional operators in the norm of that ideal. Q. E. D.

7.5 Functions of a Hilbert-Schmidt operator

Theorem 7.4. *Let A be a Hilbert-Schmidt operator and let f be a function holomorphic on a neighborhood of the closed convex hull co(A) of the spectrum of A. Then*

$$\|f(A)\| \leq \sup_{\lambda \in \sigma(A)} |f(\lambda)| + \sum_{k=1}^{\infty} \sup_{\lambda \in co(A)} |f^{(k)}(\lambda)| \frac{g^k(A)}{(k!)^{3/2}}. \tag{5.1}$$

Proof. Let a sequence $\{A_n\}$ of operators having n-dimensional ranges converges in the norm $N_2(.)$ to A. Theorem 3.5 gives us the inequality

$$\|f(A_n)\| \leq \sup_{\lambda \in \sigma(A_n)} |f(\lambda)| + \sum_{k=1}^{n-1} \sup_{\lambda \in co(A_n)} |f^{(k)}(\lambda)| \frac{g^k(A_n)}{(k!)^{3/2}}. \tag{5.2}$$

But

$$\lim_{n \to \infty} \sigma(A_n) \subseteq \sigma(A)$$

(see Section 1.6) and therefore, $\lim_{n \to \infty} co(A_n) \subseteq co(A)$. According to Lemma 1.7 (see also Lemma VII.6.5 from [18]), $\|f(A_n)\| \to \|f(A)\|$. This relation, Corollary 7.2 and (5.2) prove the theorem. Q. E. D.

Theorem 7.4 is sharp: if A is normal, inequality (5.2) becomes the equality

$$\|f(A)\| = \sup_{\mu \in \sigma(A)} |f(\mu)|,$$

because $g(A) = 0$ in this case.

Note that, if A is normal, then it is required only that f is defined on $\sigma(A)$.

Corollary 7.4. *Let A be a Hilbert-Schmidt operator. Then*

$$\|e^{At}\| \leq e^{\alpha(A)t} \sum_{k=0}^{\infty} \frac{t^k g^k(A)}{(k!)^{3/2}} \quad (t \geq 0),$$

where $\alpha(A) = \sup Re\ \sigma(A)$. In addition,

$$\|A^m\| \leq \sum_{k=0}^{m} \frac{m! r_s^{m-k}(A) g^k(A)}{(m-k)!(k!)^{3/2}} \quad (m = 1, 2, ...).$$

Recall that $r_s(A)$ is the spectral radius of an operator A. In particular, if $V \in SN_2$ is a quasi-nilpotent operator, then

$$\|V^m\| \leq \frac{N_2^m(V)}{\sqrt{m!}} \quad (m = 1, 2, ...). \tag{5.3}$$

7.6 Equations with compact operators

Let $\mathcal{E}$ be a separable Hilbert space, $A \in \mathcal{B}(\mathcal{H}), B \in \mathcal{B}(\mathcal{E})$. In addition,

$$A, B \in SN_2. \tag{6.1}$$

Again consider the equation

$$Y - AYB = C, \tag{6.2}$$

assuming that $C \in \mathcal{B}(\mathcal{E}, \mathcal{H})$ and there is a constant $b \in \mathbf{C}, b \neq 0$, such that

$$|b| r_s(A) < 1 \text{ and } r_s(B) < |b|. \tag{6.3}$$

Due to Lemma 2.2 the unique solution Y to (6.1) is representable by the integral

$$Y = \frac{b}{2\pi} \int_0^{2\pi} (Ie^{-i\omega} - bA)^{-1} C(be^{i\omega}I - B)^{-1} d\omega.$$

Obviously,

$$\rho(bA, e^{-i\omega}) \geq 1 - r_s(bA) = 1 - |b| r_s(A), \ \rho(B, be^{i\omega}) \geq |b| - r_s(B).$$

Thanks to Theorem 7.1,

$$\|(bIe^{-i\omega} - B)^{-1}\| \leq \sum_{j=0}^{\infty} \frac{g^j(B)}{\sqrt{j!}(|b| - r_s(B))^{j+1}}$$

and

$$\|(Ie^{-i\omega} - bA)^{-1})\| \leq \sum_{j=0}^{\infty} \frac{g^j(A)}{\sqrt{j!}(1 - |b|r_s(A))^{j+1}}.$$

We thus get

Corollary 7.5. *Let condition (6.1) and (6.3) hold. Then the unique solution Y of (6.2) satisfies the inequality*

$$\|Y\| \leq \|C\|\|b\| \left(\sum_{j,k=0}^{\infty} \frac{g^j(A)g^k(B)}{\sqrt{j!k!}(1 - |b|r_s(A))^{k+1}(|b| - r_s(B))^{j+1}} \right).$$

Similarly, Theorems 7.2 and 7.3 enable us to investigate equation (6.2) under the condition $A, B \in SN_{2p}$ for an integer $p > 1$.

7.7   Perturbations of compact operators

From Lemma 1.10 and (2.4) it follows

Theorem 7.5. *Let $A \in SN_2$ and $\tilde{A}$ be a bounded linear operator in $\mathcal{H}$. Then $\mathrm{sv}_A(\tilde{A}) \leq z_2(A, q)$, where $z_2(A, q)$ is the unique positive root of the equation*

$$\frac{q}{z} \exp\left[\frac{1}{2} + \frac{g^2(A)}{2z^2}\right] = 1. \tag{7.1}$$

To estimate $z_2(A, q)$ we can apply the following result.

Lemma 7.7. *For any integer $p \geq 1$, the unique positive root z_a of the equation*

$$y \exp\left[(y + 1)^p\right] = a \quad (a = const > 0) \tag{7.2}$$

satisfies the inequality $z_a \geq \delta_p(a)$, where

$$\delta_p(a) := \begin{cases} 2^{-1/p}(\ln a + 2^p)^{1/p} - 1 & \text{if } a \geq \exp\left[2^p\right], \\ \exp\left[-2^p\right]a & \text{if } a < \exp\left[2^p\right]. \end{cases}$$

Proof. First assume that $a < \exp[2^p]$. Then $z_a < 1$ and (7.2) implies

$$z_a \exp[2^p] > a.$$

So

$$z_a > \exp[-2^p]a.$$

Now let $a \geq \exp[2^p]$, then $z_a \geq 1$. Take into account that

$$f(x) := x \leq h(x) := \exp[-2^p + (x+1)^p], x \geq 1.$$

Indeed, $f(1) = h(1) = 1$ and

$$f'(x) = 1 \leq h'(x) = p(x+1)^{p-1} \exp[-2^p + (x+1)^p], x \geq 1.$$

Hence, since $z_a \geq 1$, we can write

$$z_a \leq \exp[-2^p + (z_a+1)^p]$$

and therefore (7.2) implies

$$\exp[-2^p + 2(z_a+1)^p] \geq a.$$

Consequently,

$$z_a \geq [(\ln a + 2^p)/2]^{1/p} - 1,$$

as claimed. Q. E. D.

Furthermore, with $g(A) \neq 0$ substitute the equality $z = xg(A)$ into equation (7.1). Then we arrive at the equation

$$\frac{q}{xg(A)} \exp\left[\frac{1}{2} + \frac{1}{2x^2}\right] = 1.$$

Hence

$$\frac{1}{x^2} \exp\left[1 + \frac{1}{x^2}\right] = \frac{g^2(A)}{q^2}.$$

To estimate the positive root x_0 of this equation apply Lemma 7.7 with

$$p = 1 \text{ and } a = \frac{g^2(A)}{q^2}.$$

Then we obtain the inequality

$$\frac{1}{x_0^2} \geq \delta(A, q),$$

where

$$\delta(A, q) := \begin{cases} \frac{g^2(A)}{(qe)^2} & \text{if } g(A) \leq qe, \\ \ln\left(\frac{g(A)}{q}\right) & \text{if } g(A) > qe. \end{cases}$$

Thus

$$z_2(A,q) \le g(A)x_0 \le \frac{g(A)}{\sqrt{\delta(A,q)}}.$$

Consequently, we arrive at

Corollary 7.6. *Let $A \in SN_2$ be non-normal and $\tilde{A} \in \mathcal{B}(\mathcal{H})$. Then*

$$\mathrm{sv}_A(\tilde{A}) \le \frac{g(A)}{\sqrt{\delta(A,q)}}.$$

Note that

$$\frac{g(A)}{\sqrt{\delta(A,q)}} \to qe \text{ as } g(A) \to 0.$$

So for a normal operator A we obtain $\mathrm{sv}_A(\tilde{A}) \le qe$.

Sharper but more complicated results can be obtained if instead of (2.4) by virtue of inequality (2.3). Then instead of (7.1) we obtain equation

$$q \sum_{k=0}^{\infty} \frac{g^k(A)}{z^{k+1}\sqrt{k!}} = 1. \tag{7.3}$$

From this equation and Lemma 1.10 it easily follows that $\mathrm{sv}_A(\tilde{A}) \le q$ for a normal A.

Now we are going to estimate the Hausdorff distance assuming that $A, \tilde{A} \in SN_2$. To this end put $\hat{g}(A, \tilde{A}) = \max\{g(A), g(\tilde{A})\}$. Making use of Theorem 7.5 we get our net result.

Corollary 7.7. *Let $A, \tilde{A} \in SN_2$. Then $hd(A, \tilde{A}) \le \hat{z}_2(A, \hat{A}, q)$, where $\hat{z}_2(A, \hat{A}, q)$ is the unique positive root of the equation*

$$\frac{q}{z} \exp\left[\frac{1}{2} + \frac{\hat{g}^2(A, \tilde{A})}{2z^2}\right] = 1.$$

To estimate $\hat{z}_2(A, \hat{A}, q)$ one can apply Lemma 7.7. Moreover, repeating the arguments of Section 1.9, and making use of Theorem 7.1 one can obtain a bound for the norm of difference of the simple eigenvectors of A and $\tilde{A}$ if $A \in SN_2$.

Similarly, by the estimates for the resolvent, derived above in Sections 7.3 and 7.4, one can investigate spectrum perturbations and the rotation of the simple eigenvectors of Schatten-von Neumann operators.

7.8   Comments to Chapter 7

The first time Theorems 7.1 and 7.2 appear in [24] (see also [28, Chapter 6]). They are deeply connected with the Carleman inequality, cf. [17].

Theorem 7.4 is a refinement of the norm estimate for the function of a Hilbert-Schmidt operator established in [25].

Some applications of the results presented in this chapter to the Barbashin type integro-differential equations can be found in [41].

Chapter 8

Triangular Representations of Non-selfadjoint Operators

The present chapter deals with non-compact non-normal operators A in a separable Hilbert space $\mathcal{H}$. Besides, it is supposed that either $A^* - A \in SN_p$ or $A^*A - I \in SN_p$ $(p \geq 1)$. We suggest the triangular representations of the considered operators via chains of their invariant projections. These representations are our main tool in the next two chapters devoted to operator functions of non-selfadjoint operators and the relevant operator equations.

All the operators considered in this chapter are bounded operators in $\mathcal{H}$.

8.1 $\mathcal{P}$-triangular operators

For two orthogonal projections P_1, P_2 in $\mathcal{H}$ we write $P_1 < P_2$ if $P_1\mathcal{H} \subset P_2\mathcal{H}$. A set $\mathcal{P}$ of orthogonal projections in $\mathcal{H}$ containing at least two orthogonal projections is called *a chain*, if from $P_1, P_2 \in \mathcal{P}$ with $P_1 \neq P_2$ it follows that either $P_1 < P_2$ or $P_1 > P_2$. For two chains $\mathcal{P}_1, \mathcal{P}_2$ we write $\mathcal{P}_1 < \mathcal{P}_2$ if from $P \in \mathcal{P}_1$ it follows that $P \in \mathcal{P}_2$. In this case we say that $\mathcal{P}_1$ precedes $\mathcal{P}_2$. The chain that precedes only itself is called *a maximal chain*.

Let $P^-, P^+ \in \mathcal{P}$, and $P^- < P^+$. If for every $P \in \mathcal{P}$ we have either $P < P^-$ or $P > P^+$, then the pair (P^+, P^-) is called a gap of $\mathcal{P}$. Besides, $\dim (P_+\mathcal{H}) \ominus (P_-\mathcal{H})$ is the dimension of the gap.

An orthogonal projection P in $\mathcal{H}$ is called a limit projection of a chain $\mathcal{P}$ if exists a sequence $P_k \in \mathcal{P}$ $(k = 1, 2, ...)$ which strongly converges to P. A chain is said to be closed if it contains all its limit projections.

We need the following result proved in [52, Proposition XX.4.1, p. 478], [12, Theorem II.14.1].

Theorem 8.1. *A chain is maximal if and only if it is closed, contains 0 and I, and all its gaps (if they exist) are one dimensional.*

We will say that *a maximal chain $\mathcal{P}$ is invariant for A, or A has a maximal invariant chain $\mathcal{P}$*, if $PAP = AP$ for any $P \in \mathcal{P}$.

Any compact operator has a maximal invariant chain [55, Theorem I.3.1].

Let $\sigma_d(A)$ be the discrete spectrum of A, that is, the set of all eigenvalues of A with finite algebraic multiplicities and which are isolated points of $\sigma(A)$. The essential spectrum $\sigma_{ess}(A)$ of A is defined as the complement of $\sigma_d(A)$ in $\sigma(A)$.

Definition 8.1. Let

$$A = D + V, \tag{1.1}$$

where $D \in \mathcal{B}(\mathcal{H})$ is a normal operator and V is a compact quasi-nilpotent operator in $\mathcal{H}$. Let V have a maximal invariant chain $\mathcal{P}$ and $PD = DP$ for all $P \in \mathcal{P}$. In addition, let the essential spectrum of A lie on an unclosed Jordan curve. Then A will be called *a $\mathcal{P}$-triangular operator, equality (1.1) is its triangular representation, D and V are the diagonal and nilpotent parts of A, respectively.*

We need the following lemmas.

Lemma 8.1. *Let a sequence of compact quasi-nilpotent operators $V_n \in \mathcal{B}(\mathcal{H})$ $(n = 1, 2, ...)$ converge in the operator norm to an operator V. Then V is compact quasi-nilpotent.*

Proof. We follow the proof of Lemma 17.1 from [12].

Since the uniform limit of compact operators is compact, V is compact. Assume that V has an eigenvalue $\lambda_0 \neq 0$. Since V is compact, λ_0 is an isolate point of $\sigma(V)$. So there is a circle L which contains λ_0 and does not contain zero and other points of $\sigma(V)$. We have

$$\|R_z(V_n)\| - \|R_z(V)\| \leq \|R_z(V_n) - R_z(V)\| \leq \|V - V_n\| \|R_z(V_n)\| \|R_z(V)\|.$$

Hence, for sufficiently large n,

$$\|R_z(V_n)\| \leq \frac{\|R_z(V)\|}{1 - \|V - V_n\| \|R_z(V_n)\| \|R_z(V)\|}.$$

So $\|R_z(V_n)\|$ are uniformly bounded on L. Since V_n $(n = 1, 2, ...)$ are quasi-nilpotent operators, we have

$$\int_L R_z(V_n)dz = 0$$

and

$$\int_L R_z(V)dz = \int_L (R_z(V) - R_z(V_n))dz = \int_L R_z(V)(V - V_n)R_z(V_n)dz \to 0.$$

So $\int_L R_z(V)dz = 0$ but this is impossible, since that integral represents the eigen-projection corresponding to λ_0. This contradiction proves the lemma. Q. E. D.

Lemma 8.2. *Let a compact operator $V \in \mathcal{B}(\mathcal{H})$ have a maximal invariant chain $\mathcal{P}$. If, in addition,*

$$(P^+ - P^-)V(P^+ - P^-) = 0 \qquad (1.2)$$

for every gap (P^+, P^-) of $\mathcal{P}$ (if it exists), then V is a quasi-nilpotent operator.

This result is due to Corollary XXI.1.3 from [52] (see also Corollary 1 to Theorem 17.1 of the book by M. Brodskii [12]). In particular, if $\mathcal{P}$ is continuous (that is, it does not have gaps), invariant for V, and V is compact, then V is quasi-nilpotent, provided $\mathcal{P}$ is invariant with respect to V.

We need also the following lemma.

Lemma 8.3. *Let V be a compact quasi-nilpotent operator having a maximal invariant chain $\mathcal{P}$. Then equality (1.2) holds for every gap (P^+, P^-) of $\mathcal{P}$ (if it exists).*

This result is also due to the just mentioned Corollary XXI.1.3 from [52] (see also the equality (I.3.1) from the book by [55]).

In the sequel the expression $(P^+ - P^-)T(P^+ - P^-)$ for a $T \in \mathcal{B}(\mathcal{H})$ will be called *the block of the gap (P^+, P^-) of $\mathcal{P}$ on T.*

Lemma 8.4. *Let V_1 and V_2 be compact quasi-nilpotent operators having a joint maximal invariant chain $\mathcal{P}$. Then $V_1 + V_2$ is a quasi-nilpotent operator having the same maximal invariant chain.*

Proof. Since the blocks of the gaps of $\mathcal{P}$ on both V_1 and V_2, if they exist, are zero (due to Lemma 8.3), the blocks of the gaps of $\mathcal{P}$ on $V_1 + V_2$ are also zero. Now the required result is due to Lemma 8.2. Q. E. D.

Lemma 8.5. *Let V and B be bounded linear operators in $\mathcal{H}$ having a joint maximal invariant chain $\mathcal{P}$. In addition, let V be a compact quasi-nilpotent* *operator. Then VB and BV are quasi-nilpotent, and $\mathcal{P}$ is their maximal invariant chain.*

Proof. It is obvious that

$$PVBP = VPBP = VBP \ (P \in \mathcal{P}).$$

Now let $Q = P^+ - P^-$ for a gap (P^+, P^-). Then according to Lemma 8.3 equality (1.2) holds. Further, we have $QVP^- = QBP^- = 0$,

$$QVBQ = QVB(P^+ - P^-) = QV(P^+ BP^+ - P^- BP^-) =$$

$$QV[(P^- + Q)B(P^- + Q) - P^- BP^-] = QVQBQ = 0.$$

Due to Lemma 8.2 this relation implies that VB is a quasi-nilpotent operator. Similarly we can prove that BV is quasi-nilpotent. Q. E. D.

Corollary 8.1. *Let A be $\mathcal{P}$-triangular. Let D and V be the diagonal part and nilpotent one of A, respectively. Then for any regular point λ of D, the operators $VR_\lambda(D)$ and $R_\lambda(D)V$ are quasi-nilpotent ones. Besides A, $VR_\lambda(D)$ and $R_\lambda(D)V$ have the joint maximal invariant chain.*

Indeed, we have

$$P = P(D - I\lambda)R_\lambda(D) = (D - I\lambda)PR_\lambda(D) \text{ for all } P \in \mathcal{P}.$$

Hence,

$$(D - I\lambda)^{-1}P = PR_\lambda(D).$$

Now Lemma 8.5 ensures the required result, since the nilpotent part is compact.

Lemma 8.6. *Let V and B be bounded linear operators in $\mathcal{H}$ having a joint maximal invariant chain $\mathcal{P}$. In addition, let V be a compact quasi-nilpotent operator and the regular set of B is simply connected. Then $\sigma(B + V) = \sigma(B)$.*

Proof. We have

$$PR_\lambda(B)P = -\sum_{k=0}^{\infty} P\frac{B^k}{\lambda^{k+1}}P = PR_\lambda(B)P \ (|\lambda| > \|B\|, P \in \mathcal{P}).$$

Since the set of regular points of B is simply connected, by the resolvent identity one can extend the equality $PR_\lambda(B)P = R_\lambda(B)P$ to all regular λ of B (see also [74, Corollary 2.13]).

Put $A = B + V$. For any $\lambda \notin \sigma(B)$ operator $V R_\lambda(B)$ is quasi-nilpotent due to Lemma 8.5. So $I + V R_\lambda(B)$ is boundedly invertible, and therefore,

$$R_\lambda(A) = (B + V - \lambda I)^{-1} = R_\lambda(B)(I + V R_\lambda(B))^{-1} \quad (\lambda \notin \sigma(B)). \quad (1.3)$$

Hence it follows that λ is a regular point for A. Consequently,

$$\sigma(A) \subseteq \sigma(B). \quad (1.4)$$

So the regular set of A is also simply connected.

Now let $\lambda \notin \sigma(A)$. Since $\mathcal{P}$ is invariant for A, as above we can show that $\mathcal{P}$ is invariant for $R_\lambda(A)$. Then operator $V R_\lambda(A)$ is quasi-nilpotent due to Lemma 8.5. So $I - V R_\lambda(A)$ is boundedly invertible. Furthermore, according to the equality $B = A - V$, we get

$$R_\lambda(B) = (A - V - \lambda I)^{-1} = R_\lambda(A)(I - V R_\lambda(A))^{-1}.$$

Hence, it follows that λ is a regular point also for B and therefore $\sigma(B) \subseteq \sigma(A)$. Now (1.4) proves the result. Q. E. D.

From the latter lemma it follows.

Corollary 8.2. *Let A be $\mathcal{P}$-triangular. Then $\sigma(A) = \sigma(D)$, where D is the diagonal part of A.*

Recall the Weyl inequalities [54, Lemma II.6.1].

Lemma 8.7. *Let $A \in \mathcal{B}(\mathcal{H})$ and $\Im A = (A - A^*)/2i$ be compact. Let $\lambda_j = \lambda_j(A)$ be the non-real eigenvalues of A taken with their multiplicities and enumerated as $|\Im \lambda_j| \geq |\Im \lambda_{j+1}|$ $(j = 1, 2, ...)$. Then*

$$\sum_{j=1}^{n} |\Im \lambda_j| \leq \sum_{j=1}^{n} s_j(\Im A) \quad (n = 1, 2, ...),$$

and therefore

$$\sum_{j=1}^{\infty} |\Im \lambda|^q \leq \sum_{j=1}^{\infty} s_j^q(\Im A) \quad (q \geq 1),$$

provided $\Im A \in SN_q$.

An order of a compact operator K is a number $\hat{p}(K)$ which equal to the lower bound of such p that $K \in SN_p$.

We need the following well-known result [55, Theorem III.6.1].

Lemma 8.8. *If the order $\hat{p}(\Im V)$ of the imaginary component of a quasi-nilpotent operator V is not less than one, then*

$$\hat{p}(\Re V) \leq \hat{p}(V) = \hat{p}(\Im V).$$

Theorem 8.2. *Assume that $A \in \mathcal{B}(\mathcal{H})$ has a maximal invariant chain $\mathcal{P}$ and $\Im A \in SN_p$ for some finite $p \geq 1$. Then A is a $\mathcal{P}$-triangular operator.*

Proof. Let $\mathcal{P}_n < \mathcal{P}$ $(n = 2, 3, ...)$ be a partitioning of $\mathcal{P}$ of the form

$$0 < P_1 < P_2 < ... < P_n = I \ (P_1, ..., P_n \in \mathcal{P}).$$

Obviously,

$$A = \sum_{j=1}^{n} \sum_{k=1}^{n} \Delta P_j A \Delta P_k = \sum_{k=1}^{n} \Delta P_k A \Delta P_k + \sum_{k=1}^{n} \sum_{j=1}^{k-1} \Delta P_j A \Delta P_k =$$

$$\sum_{k=1}^{n} \Delta P_k A \Delta P_k + \sum_{k=1}^{n} P_{k-1} A \Delta P_k$$

$$(\Delta P_k = P_k - P_{k-1}, k = 1, 2, ..., n).$$

But $P_{k-1} A^* \Delta P_k = 0$. So $P_{k-1} A \Delta P_k = 2i P_{k-1} \Im A \Delta P_k$. Hence,

$$A = D_n + V_n, \tag{1.5}$$

where

$$D_n = \sum_{k=1}^{n} \Delta P_k A \Delta P_k \text{ and } V_n = 2i \sum_{k=1}^{n} P_{k-1} \Im A \Delta P_k.$$

According (1.5), $\Im A = \Im D_n + \Im V_n$.

Due to the Weyl inequalities (Lemma 8.7), we can write $N_p(\Im D_n) \leq N_p(\Im A)$. Hence,

$$N_p(\Im V_n) \leq N_p(\Im A) + N_p(\Im D_n) \leq 2N_p(\Im A).$$

Furthermore, since $1 \leq p < \infty$, due to the Macaev theorem [55, Theorem III.6.2] there is a constant d_p dependent on p, only, such that $N_p(V_n) \leq d_p N_p(\Im V_n)$, and therefore,

$$N_p(V_n) \leq 2 b_p N_p(\Im A) \ (n = 1, 2, ...). \tag{1.6}$$

Thus there is a subsequence ν of natural numbers, such that the sequence $\{V_n\}_{n \in \nu}$ weakly converges (see [18, p. 512]). Its limit is denoted by V. Therefore, $\{V_n\}_{n \in \nu} \to V$ in the norm of SN_p $(p > 1)$ (for the relevant results see [18, Corollary VI.5.3]).

Now (1.5) implies that the operators D_n $(n \in \nu)$ converge to an operator $D = A - V$ in the norm of SN_p. It can be directly checked that

$$V_n^n = V_n^n P_n = V_n^{n-1} P_{n-1} V P_n = V_n^{n-2} P_{n-2} V P_{n-1} V P_n = ... = 0.$$

Hence, according to Lemma 8.1, V is quasi-nilpotent. By Lemma 8.8 we can include the case $p = 1$: if $\Im A \in SN_1$, then $V \in SN_1$.

Recall that $\Re A = (A + A^*)/2$. Since D_n $(n \in \nu)$ converge to D, the operators

$$\Re D_n = \sum_{k=1}^{n} \Delta P_k(\Re A)\Delta P_k$$

and

$$\Im D_n = \sum_{k=1}^{n} \Delta P_k(\Im A)\Delta P_k \quad (n \in \nu)$$

converge to $\Re D$ and $\Im D$, respectively; $\Re D$ and $\Im D$ are selfadjoint and $D = \Re D + i\Im D$.

Since $\Im D$ is compact and $(\Im D)P = P(\Im D)$, we can write

$$\Im D = \sum_{k=1}^{\infty} \mu_k \left(P_k^+ - P_k^-\right), \tag{1.7}$$

where (P_k^+, P_k^-) are gaps of $\mathcal{P}$ and μ_k $(k = 1, 2, ...)$ are real numbers converging to zero. Moreover,

$$P(\Re D) = (\Re D)P \quad (P \in \mathcal{P}).$$

Hence, $(\Re D)P_k^{\pm} = P_k^{\pm}(\Re D)$ and due to (1.7), $(\Re D)(\Im D) = (\Re D)(\Im D)$. Consequently, D is normal. This proves the theorem. Q. E. D.

8.2 Existence of invariant maximal chains

It is assumed that

$$\Im A \in SN_p \tag{2.1}$$

for some finite $p \geq 1$. We need the following result, cf. [74, Corollary 6.15].

Theorem 8.3. *Let condition (2.1) hold. Then A has a nontrivial invariant subspace.*

A simple corollary of this theorem is the following result.

Theorem 8.4. *If two subspaces J_1 and J_2 are invariant with respect to $A \in \mathcal{B}(\mathcal{H})$, satisfying condition (2.1), $J_1 \subset J_2$ and $\dim (J_2 \ominus J_1) > 1$, then there exists a subspace J_0 invariant for A, which is intermediate between J_1 and J_2: $J_1 \subset J_0 \subset J_2$ and $J_1 \neq J_0 \neq J_2$.*

Proof. This proof is a modification of the proof of Theorem I.2.2 from [55].

We say that a chain is bordered if it includes 0 and I. Let us denote by A_0 the operator which A induces in J_2 and by P_k the orthogonal projection onto J_k $(k = 1, 2)$. So $A_0 = AP_2$ and

$$(P_2 - P_1)A_0^*(P_2 - P_1) = P_2A^*P_2 - P_2A^*P_1 - P_1A^*P_2 + P_1A^*P_1$$

$$= P_2A^*P_2 - P_2A^*P_1 - P_1A^*P_1 + P_1A^*P_1 = P_2A^*(P_2 - P_1) = A_0^*(P_2 - P_1).$$

So the subspace $\hat{J}_1 = J_2 \ominus J_1$ is invariant with respect to A_0^*. Since

$$(P_2 - P_1)(A - A^*)(P_2 - P_1) \in SN_p,$$

according to Theorem 8.3, there is a nontrivial subspace $\hat{J}_2 \subset \hat{J}_1$ which is invariant with respect to A_0^*. The subspace $J_0 = J_2 \ominus \hat{J}_2$ satisfies the required relations and is invariant with respect to A. Indeed, let $\hat{P}_2$ be the orthogonal projection onto $\hat{J}_2$. Then $A_0^*\hat{P}_2 = \hat{P}_2A_0^*\hat{P}_2$ and therefore, $A^*\hat{P}_2 = \hat{P}_2A^*\hat{P}_2$. Hence, $\hat{P}_2A = \hat{P}_2A\hat{P}_2$. Thus,

$$(P_2 - \hat{P}_2)A(P_2 - \hat{P}_2) = P_2AP_2 - P_2A\hat{P}_2 - \hat{P}_2AP_2 + \hat{P}_2A\hat{P}_2 =$$

$$P_2AP_2 - P_2A\hat{P}_2 - \hat{P}_2A\hat{P}_2 + \hat{P}_2A\hat{P}_2 = P_2A(P_2 - \hat{P}_2) = (P_2 - I)A(P_2 - \hat{P}_2)$$

$$+A(P_2 - \hat{P}_2) = (P_2 - I)A(P_2 - I)(P_2 - \hat{P}_2) + A(P_2 - \hat{P}_2) = A(P_2 - \hat{P}_2),$$

as claimed. Q. E. D.

Theorem 8.5. *Let condition (2.1) hold. Then A has a maximal invariant chain.*

Proof. This proof is a modification of the proof of Theorem I.3.1 from [55].

We say that a chain is bordered if it includes 0 and I. Let us denote by U the set of all bordered chains invariant with respect to A. U is a partially ordered set.

If some set Z_0 of chains is linearly ordered, then, taking the union of all projectors, each of which belongs to at least one chain from Z_0, we obtain a chain which is the supremum of Z_0. Therefore by the Zorn lemma (see e.g. [18, p. 6]) there exists at least one maximal element in U. Let $\mathcal{P}$ be a fixed maximal element of the set U. The chain $\mathcal{P}$ is closed. Indeed, the closure $\overline{\mathcal{P}}$ belongs to U and $\mathcal{P} < \overline{\mathcal{P}}$ and hence $\mathcal{P} = \overline{\mathcal{P}}$.

Since the chain $\mathcal{P}$ is bordered, to complete the proof of the theorem it remains to show that all its gaps are one-dimensional. Let us assume the contrary, i.e. that some pair of projectors P_1, P_2 forms a gap of dimension greater than one. But then, by Theorem 8.4, there exists an orthogonal projection Q, such that $P_1 < Q < P_2$ and $QAQ = AQ$. This contradicts the maximality of $\mathcal{P}$. The theorem is proved. Q. E. D.

The previous theorem and Theorem 8.2 imply

Corollary 8.3. *Assume that condition (2.1) hold. Then A is a $\mathcal{P}$-triangular operator.*

8.3 Operators with real spectra

Let $\mathcal{P}$ be a maximal chain and $\mathcal{P}_n < \mathcal{P}$ ($n < \infty$) be a partitioning of $\mathcal{P}$ of the form

$$0 = P_0 < P_1 < P_2 < ... < P_n = I.$$

Let $T \in \mathcal{B}(\mathcal{H})$ and $\phi(P)$ be a scalar valued function of $P \in \mathcal{P}$. If for some operators W_1, W_2 and W_3, and any $\epsilon > 0$, there is a partitioning $\mathcal{P}_n$ of $\mathcal{P}$, such that

$$\left\| W_1 - \sum_{k=1}^{n} P_k \, T \Delta P_k \right\| < \epsilon, \quad \left\| W_2 - \sum_{k=1}^{n} \Delta P_k T \Delta P_k \right\| < \epsilon$$

and

$$\left\| W_3 - \sum_{k=1}^{n} \phi(P_k) \Delta P_k \right\| < \epsilon \quad (P_k \in \mathcal{P}_n, \Delta P_k = P_k - P_{k-1}),$$

then W_1, W_2 and W_3 are called *integrals in the Shatunovsky sense*. We write

$$W_1 = \int_{\mathcal{P}} PTdP, \quad W_2 = \int_{\mathcal{P}} dPTdP$$

and

$$W_3 = \int_{\mathcal{P}} \phi(P)dP.$$

For more details about such types integrals see [13], [52, Chapters XX and XI], and references therein. Besides, we will say that $\int_{\mathcal{P}} dPTdP$ is *the diagonal of T along $\mathcal{P}$*.

If A has a maximal invariant chain $\mathcal{P}$ and $\sigma(A)$ is real, then we will say that $\mathcal{P}$ *separates* $\sigma(A)$, if for any $t \in \sigma(A)$, there is $P_t \in \mathcal{P}$, such that

$$\sigma(AP_t/P_t\mathcal{H}) \subset (-\infty, t] \text{ and } \sigma((I - P_t)A(I - P_t)/(I - P_t)\mathcal{H}) \subset [t, \infty).$$

Here A/H_1 means the restriction of A onto a subspace $H_1 \subset \mathcal{H}$.

Furthermore, let $a(P)$ be a real valued function of $P \in \mathcal{P}$. It is said to be nondecreasing, if $a(P) \leq a(P_1)$ for $P < P_1$. It is left-continuous, if

$$\lim_{P_t \to P, P_t < P} a(P_k) = a(P) \ (P_t \in \mathcal{P}).$$

Let $S = S^* \in \mathcal{B}(\mathcal{H})$. Due to Theorem 8.5 with $\Im A = 0$, S has a maximal invariant chain $\mathcal{P}$. Moreover, $PS = SP \ (P \in \mathcal{P})$.

Lemma 8.9. *Let $S = S^* \in \mathcal{B}(\mathcal{H})$ and $\mathcal{P}$ be its maximal invariant chain. Then there is a non-decreasing real function $c(P)$ defined on $\mathcal{P}$, such that*

$$S = \int_{\mathcal{P}} c(P) dP.$$

Proof. Due to the spectral theorem for selfadjoint operators, S has a maximal chain $\mathcal{P}$ which separates $\sigma(S)$. Put

$$c(P) = \sup_{Q \leq P} \sigma(S/Q\mathcal{H}).$$

Let $\mathcal{P}_m = \{P_k\}_{k=0}^m$ be a partitioning of $\mathcal{P}$, such that $P_0 = 0, P_m = I$. In addition, P_{k-1}, P_k are either gap or

$$c(P_k) - c(P_{k-1}) < \epsilon.$$

Put $Q_k = P_k - P_{k-1}$ and $c(P_k) = c_k \ (k = 1, ..., m)$ and $c_0 = \inf \sigma(S)$. Besides, $c_m = \sup \sigma(S)$. Then $Q_k S Q_j = S Q_k Q_j = 0$, whenever $j \neq k$. Thus

$$S = \sum_{k=1}^m \sum_{j=1}^m Q_k S Q_j = \sum_{k=1}^m Q_k S Q_k = \sum_{k=1}^m S Q_k.$$

We can write

$$S = \int_{c_0}^{c_m} s \, dE(s),$$

where $E(s)$ is the spectral function of S. Since $\sup \sigma(SP_k) = c_k$ we obtain

$$SQ_k = \int_{c_{k-1}}^{c_k} s \, dE(s) \text{ and } Q_k = \int_{c_{k-1}}^{c_k} dE(s) = E(c_k) - E(c_{k-1}), k = 1, ..., m.$$

Therefore

$$c_k Q_k - SQ_k = \int_{c_{k-1}}^{c_k} (c_k - s) dE(s) \text{ and } \|c_k Q_k - SQ_k\| \leq c_k - c_{k-1}.$$

Hence,

$$S - \sum_{k=1}^m c_k Q_k = \sum_{k=1}^m (SQ_k - c_k Q_k) = \sum_{k=1}^m \int_{c_{k-1}}^{c_k} (s - c_k) dE(s).$$

Consequently, if $\mathcal{P}$ is continuous, then

$$\left\| S - \sum_{k=1}^{m} c_k Q_k \right\| \leq \max_k (c_k - c_{k-1}) \to 0 \text{ as } m \to \infty.$$

This proves the lemma when $\mathcal{P}$ is continuous. If it has gaps, the arguments can be obviously modified. Q. E. D.

Corollary 8.4. *Assume that condition (2.1) holds and $\sigma(A)$ is real. Then there is a maximal chain $\mathcal{P}$, and a nondecreasing function $a(P)$ defined on $\mathcal{P}$, such that*

$$A = \int_{\mathcal{P}} a(P)dP + V, \tag{3.1}$$

where V is quasi-nilpotent and $\mathcal{P}$ is invariant with respect to V. Moreover, $V \in SN_p$.

Indeed, (3.1) follows from the previous lemma and Corollaries 8.2 and 8.3. Besides, $\Im V = \Im A \in SN_p$. So Lemma 8.8 implies $V \in SN_p$.

Note that representation (3.1) firstly appears in [11] without the proof in the form inconvenient for us (see also [9, p. 69, formula (42)]).

Following [13], we will say that $A \in \mathcal{B}(\mathcal{H})$ having a maximal invariant chain $\mathcal{P}$ and a real spectrum belongs to a class $Z(\mathcal{P})$ if

a) $\mathcal{P}$ separates $\sigma(A)$

and

b) the diagonal of $\Im A$ along $\mathcal{P}$ is equal to zero.

Let us recall Theorem 3.2 from [13], which generalizes the main result from [11] and therefore, gives us an additional proof of (3.1).

Theorem 8.6. *Operator $A \in \mathcal{B}(\mathcal{H})$ admits the representation $A = G + \hat{V}$, where $\hat{V}$ is an operator having a maximal invariant chain $\mathcal{P}$ and the zero diagonal along $\mathcal{P}$, and G is a selfadjoint operator defined by $G = \int_{\mathcal{P}} a(P)dP$, where $a(P)$ is a left-continuous and nondecreasing function, if and only if $A \in Z(\mathcal{P})$.*

To show that (3.1) follows from Theorem 8.6, we need only to prove that under condition (2.1) operator $\hat{V}$ is compact quasi-nilpotent. To this end recall Theorem 3.1 from [13].

Theorem 8.7. *The following assertions are equivalent:*

a) V has a maximal invariant chain $\mathcal{P}$ *and the zero diagonal along* $\mathcal{P}$,

b) V is quasi-nilpotent and has a maximal invariant chain $\mathcal{P}$,

c) V admits the representation

$$V = 2i \int_{\mathcal{P}} P\,(\Im V)dP. \qquad (3.2)$$

Furthermore, the set of compact operators K satisfying

$$\sum_{k=1}^{\infty} \frac{s_k(K)}{2k-1} < \infty$$

is called *the Macaev ideal.*

As it was proved in [67] (see also [9, Theorem XX], [10] and [66]), any operator $A \in \mathcal{B}(\mathcal{H})$ with a real spectrum has a maximal invariant chain, which separates its spectrum, provided $\Im A$ belongs to the Macaev ideal. Certainly, if $\Im A \in SN_1$, then it belongs to the Macaev ideal. If $\Im A \in SN_p$ $(p > 1)$, then also

$$\sum_{k=1}^{\infty} \frac{s_k(\Im A)}{2k-1} \leq \left(\sum_{k=1}^{\infty} s_k^p(\Im A) \right)^{1/p} \left(\sum_{k=1}^{\infty} \frac{1}{(2k-1)^q} \right)^{1/q} < \infty \ \ (\frac{1}{p} + \frac{1}{q} = 1).$$

Furthermore, since $\mathcal{P}$ separates the spectrum, for any gap (P^+, P^-), operator AP^+ has an eigenvalue, say a: $AP^+f = af = aP^+f$ $(f \in \mathcal{H})$. Hence,

$$(P^+ - P^-)AP^+f = a(P^+ - P^-)f.$$

But $(P^+ - P^-)AP^- = 0$ and (P^+, P^-) is one dimensional. Thus,

$$(P^+ - P^-)A(P^+ - P^-) = (P^+ - P^-)AP^+ = a(P^+ - P^-).$$

Since the spectrum is real, a is real and therefore,

$$(P^+ - P^-)(A - A^*)(P^+ - P^-) = (a - \overline{a})(P^+ - P^-) = 0.$$

But $G = G^*$. So in Theorem 8.6, under (2.1) we have

$$\Im \hat{V} = \Im A \in SN_p \qquad (3.3)$$

and thus

$$(P^+ - P^-)\Im \hat{V}(P^+ - P^-) = 0 \qquad (3.4)$$

for any gap of $\mathcal{P}$. Due to Theorem III.4.1 from [55], conditions (3.3) and (3.4) provide the convergence of integral (3.2) with $V = \hat{V}$. Hence, by Theorem 8.7 the diagonal of $\Im A$ along $\mathcal{P}$ is equal to zero and $\hat{V}$ is a quasi-nilpotent operator. *Now Theorem 8.6 gives an additional proof of (3.1).*

For more details about the convergence of integral (3.2) see Theorem I.6.1 from [55] and Lemma II.17.1 from [12].

8.4 Operators with non-real spectra

Again assume that condition (2.1) holds, but now A may have a nonreal spectrum. In this section $\lambda_k(A)$ $(k = 1, 2, ...)$ are non-real eigenvalues of A taken with their multiplicities.

According to Theorem I.5.2 from [54], if $\Im A$ is compact, then the nonreal spectrum of A consists of no more than countable set of points which are normal (i.e. isolated and having finite multiplicities) eigenvalues.

Denote by $\mathcal{E}$ the linear closed hull of all the root vectors of A corresponding to non-real eigenvalues. Choose in each root subspace a Jordan basis. Then we obtain vectors ϕ_k for each of which either $A\phi_k = \lambda_k(A)\phi_k$, or $A\phi_k = \lambda_k(A)\phi_k + \phi_{k+1}$. Orthogonalizing the system $\{\phi_k\}$, we obtain the (orthonormal) Schur basis $\{e_k\}$ of the triangular representation:

$$Ae_k = a_{1k}e_1 + a_{2k}e_2 + ... + a_{kk}e_k \quad (k = 1, 2, ...) \tag{4.1}$$

with $a_{kk} = \lambda_k(A)$ (see [54, Section II.6]). Besides, $\mathcal{E}$ is an invariant subspace of A. Let Q be the orthogonal projection of $\mathcal{H}$ onto $\mathcal{E}$ and $C = AQ = QAQ$. So $\sigma(C)$ consists of the nonreal spectrum of A. Denote $M = (I-Q)A(I-Q)$ and $W = QA(I - Q)$. We have

$$A = (Q + (I - Q))A(Q + (I - Q)) = C + M + W,$$

since $(I-Q)AQ = (I-Q)QAQ = 0$. So on $Q\mathcal{H} \oplus (I-Q)\mathcal{H}$, A is represented by the matrix

$$A = \begin{pmatrix} C & W \\ 0 & M \end{pmatrix}.$$

Besides $\sigma(A) = \sigma(M) \cup \sigma(C)$, and $\sigma(M)$ is real.

Take into account that

$$Ce_k = Ae_k = a_{1k}e_1 + a_{2k}e_2 + ... + a_{kk}e_k = (D_C + V_C)e_k, \tag{4.2}$$

where

$$D_C e_k = a_{kk}e_k \quad (k \geq 1) \text{ and}$$

$$V_C e_k = a_{1k}e_1 + a_{2k}e_2 + ... + a_{k-1,k}e_k \quad (k \geq 2), V_C e_1 = 0. \tag{4.3}$$

In addition,

$$M - M^* = (I - Q)(A - A^*)(I - Q), C - C^* = Q(A - A^*)Q.$$

So $N_p(M - M^*) \leq N_p(A - A^*)$ and $N_p(C - C^*) \leq N_p(A - A^*)$. Hence $N_p(W - W^*) \leq N_p(A - A^*) + N_p(M - M^*) + N_p(C - C^*) < \infty$. Due to

Corollary 8.3 M has in $(I - Q)\mathcal{H}$ maximal invariant chain denoted by $\mathcal{P}_M$ and M is $\mathcal{P}_M$-triangular. So

$$M = D_M + V_M$$

where D_M is normal and V_M is compact quasi-nilpotent.

Put $D_A = D_M + D_C$ and $V_A = V_M + V_C + W$. According to (4.1), the chain $\mathcal{P}_C = \{\hat{P}_k\}_{k=1}^\infty$, where

$$\hat{P}_k = \sum_{j=1}^k (., e_k)e_k \ (k = 1, 2, ...), \hat{P}_0 = 0, \hat{P}_\infty = Q \qquad (4.4)$$

is the maximal invariant chain of C in subspace $Q\mathcal{H}$. Recall that e_k ($k = 1, 2, ...$) is the Schur basis.

Due to Corollary 8.3, A is $\mathcal{P}$-triangular. Built the chain $\mathcal{P}_A$ in the following way: any $P \in \mathcal{P}_A$ belongs to $\mathcal{P}_M \oplus \mathcal{P}_C$ and ordered as follows: if $P < Q$, then $P = \hat{P}_k$ for some $\hat{P}_k \in \mathcal{P}_C$. If $P > Q$, then $P = Q + P_M$, where $P_M \in \mathcal{P}_M$. Clearly, $\mathcal{P} = \mathcal{P}_A$ is the maximal chain of A.

Since V_M and V_C are quasi-nilpotent and mutually orthogonal, $V_M + V_C$ is quasi-nilpotent. $\mathcal{P}_A$ is invariant for $V_M + V_C$ and for W, and W is quasi-nilpotent, so due to Lemma 8.4 $V_A = V_M + V_C + W$ is quasi-nilpotent and $\mathcal{P}_A$ is its invariant chain.

From the Weyl inequalities (Lemma 8.7), it follows that $N_p(\Im D_A) \le N_p(\Im A)$ with $p \ge 1$, provided $\Im A \in SN_p$. Consequently, $N_p(\Im V_A) \le N_p(\Im A) + N_p(\Im D_A)$. Now Lemma 8.8 implies $V_A \in SN_p$.

We thus arrive at

Theorem 8.8. *Let condition (2.1) hold. Then A is $\mathcal{P}_A$-triangular, its nilpotent part $V_A \in SN_p$ and its diagonal part is representable as*

$$D_A = \int_{\mathcal{P}_M} a(P)dP + \sum_{k=1}^\infty \lambda_k(A)\Delta\hat{P}_k$$

$$(\Delta\hat{P}_k = \hat{P}_k - \hat{P}_{k-1}; \hat{P}_k \in \mathcal{P}_C, k = 1, 2, ...),$$

where $\lambda_k(A)$ are the nonreal eigenvalues with their multiplicities and $a(P)$ is a nondecreasing function of $P \in \mathcal{P}_M$.

8.5   Compactly perturbed unitary operators

Let U_0 be a unitary operator and

$$A = U_0 + K, \text{ where } K \in SN_p \text{ for some } p \in [1, \infty) . \qquad (5.1)$$

In addition, assume that

A has a regular point on the unit circle $\{z \in \mathbf{C} : |z| = 1\}$. (5.2)

From (5.1) we get

$$A^*A - I \in SN_p. \tag{5.3}$$

Theorem 8.9. *Let conditions (5.1) and (5.2) hold. Then A is $\mathcal{P}$-triangular.*

To prove this theorem we need the following lemma.

Lemma 8.10. *Let condition (5.1) hold and the operator $I - A$ be boundedly invertible. Then the operator*

$$B = i(I - A)^{-1}(I + A) \tag{5.4}$$

(Cayley's transformation of A) is bounded and satisfies the condition

$$B - B^* \in SN_p. \tag{5.5}$$

Proof. By (5.1) we have

$$B = i(I + U_0 + K)(I - U_0 - K)^{-1}$$

and

$$B^* = -i(I + U_0^* + K^*)(I - U_0^* - K^*)^{-1} =$$

$$-i(I + U_0^* + K^*)U_0 U_0^*(I - U_0^* - K^*)^{-1}$$

$$= i(I + U_0 + K_0)(I - U_0 + K_0)^{-1}.$$

Here $K_0 = K^* U_0$. Thus, $(B - B^*)/i = T_1 + T_2$, where

$$T_1 = (I + U_0)[(I - U_0 - K)^{-1} - (I - U_0 + K_0)^{-1}]$$

and

$$T_2 = K(I - U_0 - K)^{-1} - K_0(I - U_0 + K_0)^{-1}.$$

Since $K, K_0 \in SN_p$, we conclude that $T_2 \in SN_p$. It remains to prove that $T_1 \in SN_p$. Let us apply the identity

$$(I - U_0 - K)^{-1} - (I - U_0 + K_0)^{-1} =$$

$$(I - U_0 - K)^{-1}(K_0 + K)(I - U_0 + K_0)^{-1}.$$

Hence, $T_1 \in SN_p$. This completes the proof. Q. E. D.

Proof of Theorem 8.9: Without loss of generality assume that A has on the unit circle a regular point $\lambda_0 = 1$. In the other case we can consider instead of A the operator $A\lambda_0^{-1}$. By the previous lemma and Theorem 8.8 B is $\mathcal{P}$-triangular. So $B = D + V$ where D and V are the diagonal and nilpotent parts of B, respectively. The transformation inverse to (5.4) can be defined by the formula

$$A = (B - iI)(B + iI)^{-1} = (D + V - iI)(D + V + iI)^{-1}. \qquad (5.6)$$

Put

$$U = (D - iI)(D + iI)^{-1} \text{ and } W = A - U. \qquad (5.7)$$

D is normal, and consequently, U is normal. Since $PD = DP$ for any $P \in \mathcal{P}$, we have $PU = UP$. Moreover,

$$W = A - U = (B - iI)(B + iI)^{-1} - (D - iI)(D + iI)^{-1} = M_1 + M_2,$$

where

$$M_1 = (B + iI)^{-1}(B - D) = (B + iI)^{-1}V \in SN_r,$$
$$M_2 = (D - iI)((B + iI)^{-1} - (D + iI)^{-1})$$
$$= -(D - iI)((B + iI)^{-1}V(D + iI)^{-1}) \in SN_r.$$

Lemma 8.5 asserts that M_1 and M_2 are quasi-nilpotent operators having joint maximal invariant chain.

Recall that by Lemma 8.4, if V_1 and V_2 are compact quasi-nilpotent operators having a joint maximal invariant chain $\mathcal{P}$, then $V_1 + V_2$ is a quasi-nilpotent operator having the same maximal invariant chain. Thus, $W = M_1 + M_2$ is compact quasi-nilpotent. So A is a $\mathcal{P}$-triangular operator. Q. E. D.

Remark 8.1. Under conditions (5.1) and (5.5), suppose $\sigma(A)$ is unitary (i.e., it lies on the unit circle). Then, as it is well-known, [13, Theorem 4.4], [3, Theorem 2], A has a maximal invariant chain $\mathcal{P}$ of orthogonal projections and

$$A = U(I + \hat{V}),$$

where $\hat{V}$ is a compact quasi-nilpotent operator having the maximal invariant chain $\mathcal{P}$ and U is a unitary operator defined by

$$U = \int_{\mathcal{P}} e^{ia(P)}dP.$$

Here $a(P)$ is a nondecreasing function of $P \in \mathcal{P}$.

8.6 Comments to Chapter 8

The chapter contains mostly well-known results. Theorems 8.2 and 8.8, and Corollary 8.4 are probably new. About the representation of the resolvent of an operator having a maximal invariant chain see [23].

Chapter 9

Resolvents of Bounded Non-selfadjoint Operators

In the present chapter we derive norm estimates for the resolvents of bounded non-selfadjoint operators A. It is supposed that either A has a Schatten-von Neumann Hermitian component, or $A^*A - I$ is a nuclear operator. We also suggest bounds for the non-unitary eigenvalues of A, provided $A^*A - I$ is compact.

9.1 Resolvents of $\mathcal{P}$-triangular operators

Let A be a $\mathcal{P}$-triangular operator. That is,

$$A = D + V \ (\sigma(A) = \sigma(D)), \tag{1.1}$$

where $D \in \mathcal{B}(\mathcal{H})$ is a normal operator and V is a compact quasi-nilpotent operator in $\mathcal{H}$. Besides, $\mathcal{P}$ is a maximal invariant chain of V and $PD = DP$ for all $P \in \mathcal{P}$. In addition, the essential spectrum of A lie on an unclosed Jordan curve (see Definition 8.1 and Corollary 8.2). D and V are the diagonal and nilpotent parts of A, respectively.

According to (1.1),

$$R_\lambda(A) = (D + V - \lambda I)^{-1} = R_\lambda(D)(I + VR_\lambda(D))^{-1} \ (\lambda \notin \sigma(D)). \tag{1.2}$$

By Corollary 8.1 $VR_\lambda(D)$ is quasi-nilpotent. Furthermore, Lemmas 7.5 and 7.6 assert that for any quasi-nilpotent operator $W \in SN_{2p} \ (p = 1, 2, ...)$ we have

$$\|(I - W)^{-1}\| \le \sum_{m=0}^{p-1} \sum_{k=0}^{\infty} \frac{N_{2p}^{pk+m}(W)}{\sqrt{k!}} \tag{1.3}$$

and

$$\|(I - W)^{-1}\| \le \sqrt{e} \sum_{m=0}^{p-1} N_{2p}^m(W) \ exp \ [N_{2p}^{2p}(W)/2]. \tag{1.4}$$

147

Suppose that the nilpotent part of A satisfies the condition

$$V \in SN_{2p} \tag{1.5}$$

for some integer $p \geq 1$. Take into account that $\|R_\lambda(D)\| = \frac{1}{\rho(A,\lambda)}$. Recall that $\rho(A, \lambda)$ is the distance between $\lambda \in \mathbf{C}$ and the spectrum of A. Then (1.2) and (1.3) imply

$$\|R_\lambda(A)\| \leq \frac{1}{\rho(A,\lambda)} \sum_{m=0}^{p-1} \sum_{k=0}^{\infty} \frac{N_{2p}^{pk+m}(VR_\lambda(D))}{\sqrt{k!}} \quad (\lambda \notin \sigma(D)).$$

But

$$N_{2p}(VR_\lambda(D)) \leq N_{2p}(V)\|R_\lambda(D)\| = \frac{N_{2p}(V)}{\rho(D,\lambda)} = \frac{N_{2p}(V)}{\rho(A,\lambda)}. \tag{1.6}$$

So we get

$$\|R_\lambda(A)\| \leq \sum_{m=0}^{p-1} \sum_{k=0}^{\infty} \frac{N_{2p}^{pk+m}(V)}{\rho^{pk+m+1}(A,\lambda)\sqrt{k!}} \quad (\lambda \notin \sigma(A)). \tag{1.7}$$

Similarly, from (1.4) it follows

$$\|R_\lambda(A)\| \leq \sqrt{e} \sum_{m=0}^{p-1} \frac{N_{2p}^m(V)}{\rho^{m+1}(A,\lambda)} \, exp \, [\frac{N_{2p}^{2p}(V)}{2\rho^{2p}(A,\lambda)}] \quad (\lambda \notin \sigma(A)). \tag{1.8}$$

We thus have proved the following result.

Lemma 9.1. *Let $A \in \mathcal{B}(\mathcal{H})$ be a $\mathcal{P}$-triangular operator whose nilpotent part V satisfies condition (1.5). Then inequalities (1.7) and (1.8) are valid.*

9.2 Resolvents of operators with Hilbert-Schmidt Hermitian components

In this section we obtain a norm estimate for the resolvent under the conditions

$$A \in \mathcal{B}(\mathcal{H}) \text{ and } A_I = \Im A = (A - A^*)/(2i) \in SN_2. \tag{2.1}$$

To this end introduce the quantity

$$g_I(A) := \sqrt{2} \left[N_2^2(A_I) - \sum_{k=1}^{\infty} (\Im \lambda_k(A))^2 \right]^{1/2}. \tag{2.2}$$

Obviously, $g_I(A) \leq \sqrt{2} N_2(A_I)$.

Theorem 9.1. *Let conditions (2.1) hold. Then*

$$\|R_\lambda(A)\| \leq \sum_{k=0}^{\infty} \frac{g_I^k(A)}{\rho^{k+1}(A,\lambda)\sqrt{k!}} \tag{2.3}$$

and

$$\|R_\lambda(A)\| \leq \frac{\sqrt{e}}{\rho(A,\lambda)} exp\left[\frac{g_I^2(A)}{2\rho^2(A,\lambda)}\right] \quad (\lambda \notin \sigma(A)). \tag{2.4}$$

To prove this theorem we need the following two lemmas.

Lemma 9.2. *For any quasi-nilpotent Hilbert-Schmidt operator W one has*

$$2N_2^2(W_I) = 2N_2^2(W_R) = N_2^2(W) \ (W_I = \frac{1}{2i}(W - W^*), W_R = \frac{1}{2}(W + W^*)).$$

Proof. Indeed,

$$\text{trace } (W - W^*)^2 = \text{trace } (W^2 + (W^*)^2 - 2W^*W) = -2\text{trace } (W^*W),$$

since W is quasi-nilpotent. So we have $2N_2^2(W_I) = N_2^2(W)$; since iW is also quasi-nilpotent similarly we obtain $2N_2^2(W_R) = N_2^2(W)$, as claimed. Q. E. D.

Lemma 9.3. *Let an operator A satisfy the condition (2.1). Then it is a $\mathcal{P}$-triangular operator (due to Corollary 8.3). Moreover, $g_I(A) = N_2(V)$, where V is the nilpotent part of A.*

Proof. Let D be the diagonal part of A. From the triangular representation (1.1) it follows that

$$-4\text{trace } A_I^2 = \text{trace } (A - A^*)^2 = \text{trace } (D + V - D^* - V^*)^2$$

$$= \text{trace } (D - D^*)^2 + \text{trace } (V - V^*)^2 +$$

$$\text{trace } [(D - D^*)(V - V^*) + (V - V^*)(D - D^*)].$$

By Lemma 8.5, $V(D - D^*)$ and $V^*(D - D^*)$ are quasi-nilpotent. So

$$\text{trace } (D - D^*)V = \text{trace } V^*(D - D^*) = 0.$$

Hence,

$$-4\text{trace } A_I^2 = \text{trace } (D - D^*)^2 + \text{trace } (V - V^*)^2.$$

That is,

$$N_2^2(A_I) = N_2^2(V_I) + N_2^2(D_I) \quad (V_I = (V - V^*)/2i, D_I = (D - D^*)/2i).$$

Taking into account the previous lemma, we arrive at the equality

$$2N_2^2(A_I) - 2N_2^2(D_I) = N_2^2(V).$$

Recall that the nonreal spectrum of a quasi-Hermitian operator consists of isolated eigenvalues. Besides, $\sigma(A) = \sigma(D)$. Thus,

$$N^2(D_I) = \sum_{k=1}^{\infty} |\Im \lambda_k(A)|^2,$$

and we arrive at the required result. Q. E. D.

Proof of Theorem 9.1: The assertion of the theorem follows from Lemmas 9.1 and 9.3. Q. E. D.

Let us consider spectrum perturbations under condition (2.1). Let $\tilde{A} \in \mathcal{B}(\mathcal{H})$. Making use of Lemma 1.10 and (2.3), we get $\mathrm{sv}_A(\tilde{A}) \leq \hat{z}_1(A, q)$ ($q = \|A - \tilde{A}\|$), where $\hat{z}_1(A, q)$ is the unique positive root of the equation

$$q \sum_{k=0}^{\infty} \frac{g_I^k(A)}{z^{k+1}\sqrt{k!}} = 1.$$

Thus, *if A is normal then* $\mathrm{sv}_A(\tilde{A}) \leq q$.

Similarly, from Lemma 1.10 and (2.4), it follows

Corollary 9.1. *Let conditions (2.1) hold and $\tilde{A} \in \mathcal{B}(\mathcal{H})$. Then* $sv_A(\tilde{A}) \leq \hat{z}_2(A, q)$, *where $\hat{z}_2(A, q)$ is the unique positive root of the equation*

$$\frac{q}{z} \exp\left[\frac{1}{2} + \frac{g_I^2(A)}{2z^2}\right] = 1.$$

Estimating $\hat{z}_2(A, q)$ by Lemma 7.7 as in Section 7.7, in the case of non-normal operator A, we easily get

$$\hat{z}_2(A, q) \leq \frac{g_I(A)}{\sqrt{\hat{\delta}(A, q)}},$$

where

$$\hat{\delta}(A, q) := \begin{cases} \frac{g_I^2(A)}{(qe)^2} & \text{if } g_I(A) \leq qe, \\ \ln\left(\frac{g_I(A)}{q}\right) & \text{if } g_I(A) > qe. \end{cases}$$

So

$$sv_A(\tilde{A}) \leq \frac{g_I(A)}{\sqrt{\hat{\delta}(A, q)}}.$$

9.3 Auxiliary results

To derive estimates for the resolvents of operators with Schatten-von Neumann Hermitian components, in this section and in the next one we present some auxiliary results.

Recall that an operator-valued function $T(z) \in \mathcal{B}(\mathcal{H})$ is said to be holomorphic in a region G, if for any vectors $v, w \in \mathcal{H}$ the scalar function $(T(z)v, w)$ is holomorphic in G.

We need the following result called *the three lines theorem*, cf. [54, Theorem III. 13.1].

Theorem 9.2. *Let $T(z)$ be an operator function holomorphic in a strip $a \leq \Re z \leq b$. Suppose that the values of $T(z)$ belong, on the line $z = a + iy$ $(-\infty < y < \infty)$ to ideal SN_{r_1} $(1 \leq r_1 < \infty)$, and on the line $z = b + iy$ $(-\infty < y < \infty, b > a)$ to ideal SN_{r_2} $(r_1 < r_2 < \infty)$. If*

$$N_{r_1}(T(a + iy)) \leq C_1 \text{ and } N_{r_2}(T(b + iy)) \leq C_2 \ (-\infty < y < \infty),$$

then on every intermediate line $z = x + iy$ $(a < x < b; -\infty < y < \infty)$ the values of the operator function $T(z)$ belong to SN_r, where

$$\frac{1}{r} = \frac{1}{r_1} + t_x \left(\frac{1}{r_2} - \frac{1}{r_1} \right) \text{ and } t_x = \frac{x - a}{b - a}.$$

Moreover,

$$N_r(T(x + iy)) \leq C_1^{1-t_x} C_2^{t_x} \ (-\infty < y < \infty).$$

Furthermore, let Γ be a linear mapping acting from SN_p into SN_r for $p, r \geq 1$. Then Γ is said to be bounded if

$$\|\Gamma\|_{p \to r} := \sup_{X \in SN_p} \frac{N_r(\Gamma X)}{N_p(X)} < \infty.$$

In the next theorem we follow Theorem III.5.2 from [55].

Theorem 9.3. *Let there be numbers p_k and r_k $(k = 1, 2)$, such that $1 \leq p_1 < p_2 < \infty$ and $1 \leq r_1 \leq r_2 < \infty$. Let Γ be a bounded linear mapping acting from SN_{p_1} into SN_{r_1} and from SN_{p_2} into SN_{r_2}. Then for any p $(p_1 < p < p_2)$ Γ maps SN_p into SN_r, where*

$$\frac{1}{r} = \frac{t_p}{r_2} + \frac{1 - t_p}{r_1} \text{ and } t_p = \frac{p - p_1}{p_2 - p_1}.$$

Moreover,

$$\|\Gamma\|_{p \to r} \leq \|\Gamma\|_{p_1 \to r_1}^{1-t_p} \|\Gamma\|_{p_2 \to r_2}^{t_p}.$$

Proof. Let $X \in SN_p$ and $X = UH$ its polar decomposition, where H is a selfadjoint positive definite operator and U is a partially isometric operator. So

$$H = \sum_{j=1}^{\infty} \lambda_j(., \phi_j)\phi_j,$$

where ϕ_j are the normed eigenvectors of H.

For any $z \in \mathbf{C}$ with $\Re z \geq 0$ the operator-function H^z is defined as

$$H^z = \sum_{j=1}^{\infty} \lambda_j^z(., \phi_j)\phi_j$$

$$(\lambda^z = exp(z \ \ln \ \lambda), \ \lambda > 0, -\infty < \ln \ \lambda < \infty).$$

Consider the holomorphic operator function

$$T(z) = \Gamma(UH^z) \ \ (\Re z \geq 0).$$

If $\Re \ z = p_2/p(= a)$, then $UH^z \in SN_{p_2}$. Indeed, since $H \in SN_p$ we have

$$N_p^p(H) = \sum_{j=1}^{\infty} \lambda_j^p < \infty,$$

and therefore,

$$N_{p_2}^{p_2}(H) = \sum_{j=1}^{\infty} (\lambda_j^p)^{p_2/p} < \infty.$$

Consequently,

$$N_{r_2}(T(a+iy)) \leq \|\Gamma\|_{p_2 \to r_2} N_{p_2}(H^z) =$$

$$\|\Gamma\|_{p_2 \to r_2} N_{p_2}(H^a) = \|\Gamma\|_{p_2 \to r_2} N_p^a(H) \ \ (-\infty < y < \infty).$$

Similarly, let $\Re \ z = p_1/p(= b)$, then $UH^z \in SN_{p_1}$ and therefore,

$$N_{r_1}(T(b+iy)) \leq \|\Gamma\|_{p_1 \to r_1} N_{p_1}(H^b) = \|\Gamma\|_{p_1 \to r_1} N_p^b(H)$$

$$(-\infty < y < \infty).$$

According to the previous theorem

$$N_r(\Gamma X) = N_r(T(1)) \leq \|\Gamma\|_{p_1 \to r_1}^{1-t_p} \|\Gamma\|_{p_2 \to r_2}^{t_p} N_p^{b(1-t_p)+at_p}(H).$$

But

$$p = p_1(1-t_p) + p_2 t_p \ \text{and} \ b(1-t_p) + at_p = \frac{1}{p}(p_1(1-t_p) + p_2 t_p) = 1.$$

Thus,

$$N_r(\Gamma X) \leq \|\Gamma\|_{p_1 \to r_1}^{1-t_p} \|\Gamma\|_{p_2 \to r_2}^{t_p} N_p(H).$$

The theorem is proved. Q. E. D.

We are going to consider the case

$$p_1 = r_1 = 2^n, p_2 = r_2 = 2^{n+1}, p = 2^n c \quad (n = 1, 2, \ldots; 1 \leq c \leq 2).$$

Then due to Theorem 9.3, with

$$t_p = \frac{p - p_1}{p_2 - p_1} = \frac{2^n c - 2^n}{2^n} = c - 1,$$

we have

$$\frac{1}{r} = \frac{t_p}{p_2} + \frac{1 - t_p}{p_1} = \frac{c-1}{2^{n+1}} + \frac{2-c}{2^n}.$$

Hence

$$\frac{1}{d} = \frac{1}{2}(3 - c)$$

where $d = 2^{-n}r$. Therefore,

$$\frac{c}{d} = f(c) := \frac{1}{2}(3c - c^2).$$

The local extremum of function $f(c)$ for $1 \leq c \leq 2$ is unique, it is at $c = 3/2$ and $f(3/2) = \frac{9}{8}$. Since $f(1) = f(2) = 1$, we have

$$1 \leq \frac{c}{d} = f(c) \leq \frac{9}{8}.$$

Hence $r = 2^n d \leq 2^n c = p$ and therefore, $N_p(\Gamma X) \leq N_r(\Gamma X)$.

Now the previous theorem yields.

Corollary 9.2. *Let* $2^n \leq p \leq 2^{n+1}$ *for some integer* $n \geq 1$. *Let* Γ *be a bounded linear mapping acting from* SN_{2^n} *into itself and from* $SN_{2^{n+1}}$ *into itself. Then* Γ *maps* SN_p *into* SN_r *with* $r \leq p$ *and*

$$\|\Gamma\|_{p \to p} \leq \|\Gamma\|_{2^n \to 2^n}^{1-t_p} \|\Gamma\|_{2^{n+1} \to 2^{n+1}}^{t_p}$$

with $t_p = p2^{-n} - 1$.

9.4   Some properties of quasi-nilpotent Schatten-von Neumann operators

Our aim in the present section is to prove the following (Macaev's) theorem, cf. Theorem III.6.2 from [55].

Theorem 9.4. *Let $2 \leq p < \infty$ and $V \in \mathcal{B}(\mathcal{H})$ be a quasi-nilpotent operator with $V_I = (V - V^*)/2i \in SN_p$. Then $V_R = (V + V^*)/2 \in SN_p$ and*

$$N_p(V_R) \leq b_p N_p(V_I), \tag{4.1}$$

where the constant b_p depends on p, only.

The proof of this theorem is based on the next lemma.

Lemma 9.4. *Let $\{c_n\}_{n=1}^{\infty}$ be the sequence of the positive numbers defined by*

$$c_n = c_{n-1} + \sqrt{c_{n-1}^2 + 1} \ (n = 2, 3, ...), c_1 = 1.$$

Let V be a quasi-nilpotent operator, such that $V_I \in SN_{2^n}$ for an integer $n \geq 1$. Then $V_R \in SN_{2^n}$ and

$$N_{2^n}(V_R) \leq c_n N_{2^n}(V_I).$$

Proof. To apply the mathematical induction method assume that for $p = 2^n$ there is a constant d_p, such that $N_p(W_R) \leq d_p N_p(W_I)$ for any quasi-nilpotent operator $W \in SN_p$. Then replacing W by Wi we have $N_p(W_I) \leq d_p N_p(W_R)$. Now let $V \in S_{2p}$. Then $V^2 \in SN_p$ and therefore,

$$N_p((V^2)_R) \leq d_p N_p((V^2)_I).$$

Here

$$(V^2)_R = (V^2 + (V^2)^*)/2, (V^2)_I = (V^2 - (V^2)^*)/2i.$$

But

$$(V^2)_R = (V_R)^2 - (V_I)^2, (V^2)_I = V_I V_R + V_R V_I$$

and thus

$$N_p(V_R^2) - N_p(V_I^2) \leq N_p(V_R^2 - V_I^2) \leq d_p N_p(V_R V_I + V_I V_R) \leq$$

$$2d_p N_{2p}(V_R) N_{2p}(V_I).$$

Take into account that

$$N_p((V_R)^2) = N_{2p}^2(V_R), N_p((V_I)^2) = N_{2p}^2(V_I).$$

So

$$N_{2p}^2(V_R) - N_{2p}^2(V_I) - 2d_p N_{2p}(V_R) N_{2p}(V_I) \le 0.$$

Solving this inequality with respect to $N_{2p}(V_R)$, we get

$$N_{2p}(V_R) \le N_{2p}(V_I)[d_p + \sqrt{d_p^2 + 1}] = N_{2p}(V_I) d_{2p}$$

with

$$d_{2p} = d_p + \sqrt{d_p^2 + 1}.$$

Besides $d_2 = 1$ according to Lemma 9.2. We thus have the required result with $d_{2^n} = c_n$. Q. E. D.

Proof of Theorem 9.4: Let Γ be a mapping defined by $V_R = \Gamma V_I$. Due to the previous lemma $\|\Gamma\|_{2^n} \le c_n$ for all integer $n \ge 1$. So for any $p \in [2^n, 2^{n+1}]$, due to Corollary 9.2,

$$\|\Gamma\|_{p \to p} \le \|\Gamma\|_{2^n \to 2^n}^{1-t_p} \|\Gamma\|_{2^{n+1} \to 2^{n+1}}^{t_p} \le c_n^{1-t_p} c_{n+1}^{t_p},$$

with

$$t_p = \frac{p - 2^n}{2^n}.$$

Therefore, the required inequality holds with

$$b_p = c_n^{1-t_p} c_{n+1}^{t_p} \quad (p \in [2^n, 2^{n+1}]). \tag{4.2}$$

The theorem is proved. Q. E. D.

Furthermore, we have

$$c_{n+1} = c_n + \sqrt{c_n^2 + 1} \ge 2c_n \ge 2^n c_1 = 2^n \quad (n = 1, 2, ...).$$

Therefore, $c_n \ge 2^{n-1}$ and

$$c_{n+1} = c_n(1 + \sqrt{1 + 1/c_n^2}) \le c_n(1 + \sqrt{1 + 2^{-2(n-1)}}).$$

Hence,

$$c_{n+1} \le \prod_{k=1}^{n} (1 + \sqrt{1 + 4^{-(k-1)}}) \quad (n = 2, 3, ...). \tag{4.3}$$

Since

$$\sqrt{1 + x} \le 1 + x/2, x \in (0, 1),$$

we can write

$$1 + \sqrt{1 + 4^{-(k-1)}} \le 2(1 + 4^{-k}).$$

But $1 + x \leq e^x$ $(x \geq 0)$, and

$$\sum_{k=1}^{\infty} \frac{1}{4^k} = 1/3.$$

Consequently, from inequality (4.3) it follows that

$$c_{n+1} \leq 2^n \prod_{k=1}^{n} (1 + 4^{-k}) \leq 2^n \, e^{1/3} = 2^{n+1} \, e^{1/3}/2.$$

For $p = (1 - t)2^n + t2^{n+1}$ $(n = 1, 2, ...; \ 0 \leq t \leq 1)$ we have

$$b_p = c_n^{1-t} c_{n+1}^t \leq 2^{n(1-t)} 2^{t(n+1)} \, \frac{e^{1/3}}{2} = 2^{n+t} \, \frac{e^{1/3}}{2}.$$

Simple calculations show that

$$\max_{0 \leq t \leq 1} \frac{2^{n+t}}{p} = \max_{0 \leq t \leq 1} \frac{2^{n+t}}{(1-t)2^n + t2^{n+1}} = \max_{0 \leq t \leq 1} \frac{2^t}{1+t} = 1.$$

Hence

$$b_p \leq \frac{p}{2} e^{1/3} < p \ (2 \leq p < \infty). \tag{4.4}$$

We thus have proved

Lemma 9.5. *The constant b_p in Theorem 9.4 satisfies inequality (4.4).*

9.5 Resolvent of operators with Schatten-von Neumann Hermitian components

Let us derive a norm estimate for the resolvent under the condition

$$A_I = (A - A^*)/2i \in SN_{2p} \text{ for an integer } p \geq 2. \tag{5.1}$$

Put

$$\tau_p(A) = (1 + b_{2p})(N_{2p}(A_I) + N_{2p}(D_I)),$$

where b_{2p} is taken from (4.1). So according to Lemma 9.5,

$$\tau_p(A) \leq (1 + pe^{1/3})(N_{2p}(A_I) + N_{2p}(D_I)) \leq (1 + 2p)(N_{2p}(A_I) + N_{2p}(D_I)). \tag{5.2}$$

From the Weyl inequalities (Lemma 8.7) we have $N_{2p}(D_I) \leq N_{2p}(A_I)$. Thus in the general case

$$\tau_p(A) \leq 2(1 + 2p)N_{2p}(A_I). \tag{5.3}$$

If A has a real spectrum, then

$$\tau_p(A) \le (1 + 2p)N_{2p}(A_I). \tag{5.4}$$

Theorem 9.5. *Let $A \in \mathcal{B}(\mathcal{H})$ satisfy condition (5.1). Then*

$$\|R_\lambda(A)\| \le \sum_{m=0}^{p-1} \sum_{k=0}^{\infty} \frac{\tau_p^{pk+m}(A)}{\rho^{pk+m+1}(A, \lambda)\sqrt{k!}} \tag{5.5}$$

and

$$\|R_\lambda(A)\| \le \sqrt{e} \sum_{m=0}^{p-1} \frac{\tau_p^m(A)}{\rho^{m+1}(A, \lambda)} \exp\left[\frac{\tau_p^{2p}(A)}{2\rho^{2p}(A, \lambda)}\right] \quad (\lambda \notin \sigma(A)). \tag{5.6}$$

Proof. Theorem 9.4 implies

$$N_{2p}(V) \le N_{2p}(V_I) + N_{2p}(V_R) \le (1 + b_{2p})N_{2p}(V_I).$$

Hence, due to the triangular representation (1.1) it follows

$$N_{2p}(V) \le (1 + b_{2p})(N_{2p}(A_I) + N_{2p}(D_I)). \tag{5.7}$$

Now Lemma 9.1 implies the required result. Q. E. D.

Theorem 9.5 is sharp: if A is selfadjoint, then inequality (5.5) takes the form $\|R_\lambda(A)\| = \frac{1}{\rho(A,\lambda)}$.

Remark 9.1. If a unitary operator U commutes with A and

$$(UA)_I := (UA - (UA)^*)/2i \in SN_{2p}, \tag{5.8}$$

then repeating the proof of Theorem 9.5 one can obtain a norm estimate for the resolvent under condition (5.8). In particular, one can take an operator U defined by the multiplication by e^{it} for a real t. Then condition (5.8) takes the form

$$(e^{it}A)_I = (e^{it}A - e^{-it}A^*)/2i \in SN_{2p}.$$

9.6 Resolvents of operators close to unitary ones

Let U_0 be a unitary operator. In this section it is assumed that

$$A = U_0 + K \in \mathcal{B}(\mathcal{H}) \text{ with } K \in SN_1 \tag{6.1a}$$

and

$$A \text{ has a regular point on the unit circle } \{z \in \mathbf{C} : |z| = 1\}. \tag{6.1b}$$

So

$$AA^* - I \in SN_1. \tag{6.2}$$

Put

$$\vartheta(A) := [\text{trace } (A^*A - I) - \sum_{k=1}^{\infty}(|\lambda_k(A)|^2 - 1)]^{1/2}.$$

Here $\lambda_k(A)$ $(k = 1, 2, ...)$, are the non-unitary eigenvalues with their multiplicities (that is, the eigenvalues with the property $|\lambda_k(A)| \neq 1$). Due to Theorem 8.9 A is $\mathcal{P}$-triangular.

Lemma 9.6. *Let conditions (6.1a) and (6.1b) hold, and V be the nilpotent part of A. Then*

$$N_2(V) = \vartheta(A).$$

Proof. Employing the triangular representation (1.1), we can write

$$\text{trace } (A^*A - I) = \text{trace } [(D + V)^*(D + V) - I].$$

But D^*V is a quasi-nilpotent operator and $V \in SN_1$. So trace $(V^*D) = 0$ and

$$\text{trace } (A^*A - I) = \text{trace } (D^*D - I) + \text{trace } (V^*V).$$

Since D is a normal operator and the spectra of D and A coincide, we get

$$\text{trace } (D^*D - I) = \sum_{k=1}^{\infty} (|\lambda_k(A)|^2 - 1).$$

This equality implies the required result. Q. E. D.

Theorem 9.6. *Let conditions (6.1a) and (6.1b) hold. Then*

$$\|R_\lambda(A)\| \leq \sum_{k=0}^{\infty} \frac{\vartheta^k(A)}{\sqrt{k!}\rho^{k+1}(A, \lambda)} \quad (\lambda \notin \sigma(A)) \tag{6.3}$$

and

$$\|R_\lambda(A)\| \leq \frac{\sqrt{e}}{\rho(A, \lambda)} \exp\left[\frac{\vartheta^2(A)}{2\rho^2(A, \lambda)}\right] \quad (\lambda \notin \sigma(A)). \tag{6.4}$$

The assertion of Theorem 9.6 follows from Lemmas 9.1 and 9.6.
Q. E. D.

Inequality (6.3) becomes the equality when A is a unitary operator. It is sharper than (6.4) but more cumbersome.

Furthermore, for two selfadjoint operators S and T we write $S \geq T$ if $(Sx, x) \geq (Tx, x)$ $(x \in \mathcal{H})$.

Lemma 9.7. *Under the conditions (6.2) and*

$$A^*A \geq I, \tag{6.5}$$

one has

$$\vartheta^2(A) \leq \text{trace } (A^*A - I).$$

Proof. From (6.5) for any eigenvalue λ of A we have

$$|\lambda|^2(e, e) = (Ae, Ae) \geq (e, e) = 1,$$

where e is the normed eigenvector. Hence,

$$\text{trace } (D^*D - I) = \sum_{k=1}^{n} (|\lambda_k(A)|^2 - 1) \geq 0.$$

Thus the lemma is valid. Q. E. D.

Now, instead of (6.5) suppose that

$$A^*A \leq I \text{ and } A \text{ is invertible.} \tag{6.6}$$

Theorem 9.7. *Let conditions (6.1a) and (6.6) hold. Then*

$$\|(A - \lambda I)^{-1}\| \leq \|A^{-1}\| \frac{r_s(A)\sqrt{e}}{\rho(A, \lambda)} \exp \left[\frac{|\lambda|^2 \hat{\zeta}(A)}{2\rho^2(A, \lambda)} \right] \quad (\lambda \notin \sigma(A)), \tag{6.7}$$

where

$$\hat{\zeta}(A) = r_s^2(A)\|A^{-1}\|^2 \text{trace } (I - AA^*).$$

Proof. We have $A^{-1}(A^*)^{-1} \geq I$ and

$$(A^*)^{-1}A^{-1} - I = (A^*)^{-1}(I - A^*A)A^{-1} \in SN_1.$$

Take into account that

$$\text{trace } (A^{-1}(A^*)^{-1} - I) = N_1(A^{-1}(A^*)^{-1} - I) \leq \zeta(A),$$

where

$$\zeta(A) = \|(A^*)^{-1}\|\|A^{-1}\|N_1(AA^* - I) = \|A^{-1}\|^2 \text{trace } (I - AA^*).$$

Now the previous lemma and inequality (6.4) imply

$$\|(A^{-1} - I\lambda^{-1})^{-1}\| \le \frac{\sqrt{e}}{\rho(A^{-1}, \lambda^{-1})} \exp\left[\frac{\zeta(A)}{2\rho^2(A^{-1}, \lambda^{-1})}\right]$$

$$(\lambda \notin \sigma(A) \cup \{0\}).$$

Besides,

$$\rho(A^{-1}, \lambda^{-1}) = \inf_s |s^{-1} - \lambda^{-1}| = \inf_s |s - \lambda||s^{-1}\lambda^{-1}| \ge \frac{\rho(A, \lambda)}{r_s(A)|\lambda|}.$$

Consequently,

$$\|(A - \lambda I)^{-1}\| \le \|(A^{-1} - \lambda^{-1}I)^{-1}\|\|A^{-1}\||\lambda|^{-1} \le$$

$$\|A^{-1}\|\frac{r_s(A)\sqrt{e}}{\rho(A, \lambda)} \exp\left[\frac{|\lambda|^2 r_s^2(A)\zeta(A)}{2\rho^2(A, \lambda)}\right] \quad (\lambda \notin \sigma(A)).$$

This proves the theorem. Q. E. D.

Furthermore, let $s \in \sigma(A)$ be such that $|\lambda - s| = \rho(A, \lambda)$ for given λ. Then

$$|\lambda| \le |\lambda - s| + |s| \le \rho(A, \lambda) + r_s(A).$$

Now the preceding theorem implies

Corollary 9.3. *Under the hypothesis of Theorem 9.7 one has*

$$\|(A - \lambda I)^{-1}\| \le$$

$$\|A^{-1}\|\frac{r_s(A)}{\rho(A, \lambda)} \exp\left[\frac{1}{2} + \hat{\zeta}(A)\left(1 + \frac{r_s^2(A)}{\rho^2(A, \lambda)}\right)\right] \quad (\lambda \notin \sigma(A)).$$

The expression in the right-hand part of the inequality in Corollary 9.3 is a function decreasing with respect to $\rho(A, \lambda)$.

9.7 Eigenvalues of compactly perturbed unitary operators

Consider the operator

$$A = U + K \tag{7.1}$$

where U is a unitary operator having a regular point on the unit circle $\{z \in \mathbf{C} : |z| = 1\}$ and K is compact. So $A^*A - I$ is compact.

We derive inequalities for the sums of absolute values of the non-unitary eigenvalues.

Due to Theorem I.5.1 from [54] (see also Section 8.4) and the Caley transform, under condition (7.1) any point λ, such that $|\lambda| \neq 1$ is either regular or a non-unitary eigenvalue of A. About other similar results see [60, p. 244].

Denote by $s_k(A^*A - I)$ $(k = 1, 2, ...)$ the singular values of $A^*A - I$ enumerated in the non-increasing order.

9.7.1 *Eigenvalues outside the unit circle*

In this subsection it is supposed that all the non-unitary eigenvalues of A lie outside the unit circle. Denote by $\lambda_k(A)$ these eigenvalues enumerated in the non-increasing order of their modulus: $|\lambda_k(A)| \geq |\lambda_{k+1}(A)|$ $(k = 1, 2, ...)$. So

$$|\lambda_k(A)| \geq 1 \quad (k = 1, 2, ...). \tag{7.2}$$

Condition (7.2) is provided by the inequality $A^*A \geq I$. Clearly,

$$\sum_{k=1}^{j} |\lambda_k(A)|^2 \leq \sum_{k=1}^{j} s_k^2(A) \quad (j = 1, 2, ...)$$

and

$$\sum_{k=1}^{j} (s_k^2(A) - 1) = \sum_{k=1}^{j} (\lambda_k(A^*A) - 1) = \sum_{k=1}^{j} \lambda_k(A^*A - I)$$

$$= \sum_{k=1}^{j} s_k(A^*A - I).$$

We thus arrive at

Lemma 9.8. *Let conditions (7.1) and (7.2) hold. Then*

$$\sum_{k=1}^{j} (|\lambda_k(A)|^2 - 1) \leq \sum_{k=1}^{j} s_k(A^*A - I) \quad (j = 1, 2, ...). \tag{7.3}$$

From Lemma 9.8 and the classical Lemma II.3.4 from [54] it follows

$$\sum_{k=1}^{j} h(|\lambda_k(A)|^2 - 1) \le \sum_{k=1}^{j} h(s_k(A^*A - I)) \quad (j = 1, 2, ...)$$

for any convex function $h(x)$ $(x \ge 0)$ satisfying $h(0) = 0$. In particular,

$$\sum_{k=1}^{j} (|\lambda_k(A)|^2 - 1)^p \le \sum_{k=1}^{j} s_k^p(A^*A - I) \quad (j = 1, 2, ...)$$

for an arbitrary $p \ge 1$. If $A^*A - I \in SN_p$ $(p \ge 1)$, then under (7.2) we have

$$\sum_{k=1}^{\infty} (|\lambda_k(A)|^2 - 1)^p \le N_p^p(A^*A - I). \tag{7.4}$$

9.7.2 *Eigenvalues inside the unit circle*

Throughout this subsection it is supposed that A is invertible and all the non-unitary eigenvalues of A lie inside the unit circle. Denote by $\hat{\lambda}_k(A)$ these eigenvalues enumerated in the non-decreasing order of their modulus: $|\hat{\lambda}_k(A)| \le |\hat{\lambda}_{k+1}(A)|$ $(k = 1, 2, ...)$. So

$$|\hat{\lambda}_k(A)| \le 1 \quad (k = 1, 2, ...). \tag{7.5}$$

Condition (7.5) is provided by the inequality $A^*A \le I$. Clearly, $\lambda_k(A^{-1}) = 1/\hat{\lambda}_k(A)$ and therefore $|\lambda_k(A^{-1})| = 1/|\hat{\lambda}_k(A)| > 1$. Here $\lambda_k(A^{-1})$ are the non-unitary eigenvalues of A^{-1} enumerated in the non-increasing order. Note that according to (7.1), $A^{-1} = (U + K)^{-1} = U^{-1} + K_1$, where

$$K_1 = (U + K)^{-1} - U^{-1} = -U^{-1}KA^{-1}$$

is compact. So A^{-1} has the form (7.1). Besides,

$$A^{-1}(A^*)^{-1} - I = A^{-1}(A^*)^{-1}(I - A^*A)$$

is compact. Due to Lemma 9.8

$$\sum_{k=1}^{j} (|\lambda_k(A^{-1})|^2 - 1) \le \sum_{k=1}^{j} s_k(A^{-1}(A^{-1})^* - I).$$

Here $s_k(A^{-1}(A^{-1})^* - I)$ are the singular values of $A^{-1}(A^{-1})^* - I$, enumerated in the non-increasing order. Since

$$s_k(A^{-1}(A^*)^{-1} - I) = s_k(A^{-1}(A^*)^{-1}(I - A^*A)) \le \|A^{-1}\|^2 s_k(AA^* - I),$$

we obtain

$$\sum_{k=1}^{j} \left(\frac{1}{|\hat{\lambda}_k(A)|^2} - 1 \right) = \sum_{k=1}^{j} \frac{1 - |\hat{\lambda}_k(A)|^2}{|\hat{\lambda}_k(A)|^2} \le \|A^{-1}\|^2 \sum_{k=1}^{j} s_k(A^*A - I).$$

Hence for all integer $j \geq 1$ we get

$$\sum_{k=1}^{j}(1 - |\hat{\lambda}_k(A)|^2) \leq |\hat{\lambda}_j(A)|^2 \|A^{-1}\|^2 \sum_{k=1}^{j} s_k(AA^* - I) \leq$$

$$r_s^2(A)\|A^{-1}\|^2 \sum_{k=1}^{j} s_k(A^*A - I). \qquad (7.6)$$

We thus arrive at

Lemma 9.9. *Let and A be representable by (7.1) and invertible. Let all the non-unitary eigenvalues of A be inside the unit circle. Then inequalities (7.6) are valid, and therefore,*

$$\sum_{k=1}^{\infty}(1 - |\hat{\lambda}_k(A)|^2)^p \leq r_s^{2p}(A)\|A^{-1}\|^{2p} N_p^p(AA^* - I) \quad (p \geq 1),$$

provided $A^*A - I \in SN_p$.

9.7.3 The general case

In this subsection operator A can have the non-unitary eigenvalues $\lambda_k(A)$ $(k = 1, 2, ...)$ (taken with the multiplicities) inside and outside the unit circle.

Theorem 9.8. *Let A be invertible and $A^*A - I \in SN_p$ for an integer $p \geq 1$. Then*

$$\sum_{k=1}^{\infty}|1 - |\lambda_k(A)|^2|^p \leq (1 + r_s^{2p}(A)\|A^{-1}\|^{2p})N_p^p(AA^* - I). \qquad (7.7)$$

Proof. Let $\overline{P}$ and P be the orthogonal invariant projections of A, corresponding to the eigenvalues inside and outside the unit circle, respectively, such that $|\lambda_k(A\overline{P})| < 1$ and $|\lambda_k(AP)| > 1$ $(k = 1, 2, ...)$. In addition,

$$\sum_{k=1}^{\infty}|1 - |\lambda_k(A)|^2|^p = \sum_{j=1}^{\infty}(|\lambda_j(AP)|^2 - 1)^p + \sum_{j=1}^{\infty}(1 - |\lambda_j(A\overline{P})|^2)^p. \qquad (7.8)$$

Since $\overline{P}$ is invariant for A and A is invertible, $A\overline{P}$ is invertible in $\overline{P}\mathcal{H}$. From (7.4) and Lemma 9.9 it follows

$$\sum_{k=1}^{\infty}|1 - |\lambda_k(A)|^2|^p \leq$$

$$N_p^p(P(A^*A - I)P) + r_s^{2p}(A\overline{P})\|(A\overline{P})^{-1}\|^{2p}N_p^p(\overline{P}(A^*A - I)\overline{P}). \qquad (7.9)$$

Here $(A\overline{P})^{-1}$ is understood as the inverse of $A\overline{P}$ in $\overline{P}\mathcal{H}$. We can write

$$\|(A\overline{P})^{-1}\| = \frac{1}{\|A\overline{P}\|_{\text{low}}}$$

where

$$\|A\overline{P}\|_{\text{low}} = \inf_{x \in \overline{P}\mathcal{H}} \|A\overline{P}x\|/\|x\| \geq \inf_{y \in \mathcal{H}} \|Ay\|/\|y\|.$$

So

$$\|(A\overline{P})^{-1}\| \leq \sup_{y \in \mathcal{H}} \frac{\|y\|}{\|Ay\|} = \|A^{-1}\|.$$

Take into account that $\overline{P}$ is an invariant projection and therefore $r_s(A\overline{P}) \leq r_s(A)$. So from (7.9) it follows

$$\sum_{j=1}^{\infty} |1 - |\lambda_j(A)|^2|^p \leq$$

$$N_p^p(P(A^*A - I)P) + r_s^{2p}(A)\|A^{-1}\|^{2p}N_p^p(\overline{P}(A^*A - I)\overline{P})).$$

This proves the theorem. Q. E. D.

9.8   Additional estimates for $\vartheta(A)$

In this section we again suppose that condition (6.2) holds. Recall that $\vartheta(A)$ is defined in Section 9.6. The calculations of $\vartheta(A)$ is not an easy task, in general. We are going to apply our results from the previous section to get additional estimates for $\vartheta(A)$.

Lemma 9.10. *Under condition (6.2), let A be invertible. Then*

$$\vartheta(A) \leq \sqrt{N_1(A^*A - I)} \begin{cases} 1 & \text{if } A^*A \geq I, \\ (r_s^2(A)\|A^{-1}\|^2 - 1)^{1/2} & \text{if } A^*A < I, \\ (r_s^2(A)\|A^{-1}\|^2 + 2)^{1/2} & \text{otherwise.} \end{cases}$$

Proof. If the condition $A^*A \geq I$ holds, then due to Lemma 9.7,

$$\vartheta(A) \leq [\text{trace } (A^*A - I)]^{1/2} = N_1^{1/2}(A^*A - I).$$

If $A^*A < I$, then $|\lambda_k(A)| < 1$ and trace $(A^*A - I) = -N_1(A^*A - I)$. Due to Lemma 9.9

$$\vartheta^2(A) = -N_1(A^*A - I) + \sum_{k=1}^{\infty}(1 - |\hat{\lambda}_k(A)|^2)$$

$$\leq (r_s^2(A)\|A^{-1}\|^2 - 1)N_1(AA^* - I).$$

So in the case $A^*A < I$ the lemma is also proved. In the general case, due to Theorem 9.8,

$$\vartheta^2(A) \leq N_1(A^*A - I) + \sum_{k=1}^{\infty} |1 - |\tilde{\lambda}_k(A)|^2| \leq$$

$$(2 + r_s^2(A)\|A^{-1}\|^2)N_1(AA^* - I).$$

This finishes the proof. Q. E. D.

9.9 Multiplicative representations of resolvents

Let F be a function defined on a maximal chain $\mathcal{P}$ with values in $\mathcal{B}(\mathcal{H})$ and $\mathcal{P}_n = \{0 = P_0 < P_1 < ... < P_n = I\}$ be a partitioning of $\mathcal{P}$. Put

$$M_n(\mathcal{P}) := \overset{\rightarrow}{\prod_{1 \leq k \leq n}} (1 + \Delta F(P_k)) := (1 + \Delta F(P_1))(I + \Delta F(P_2))...(I + \Delta F(P_n)).$$

Here $\Delta F(P_k) = F(P_k) - F(P_{k-1})$ ($P_k \in \mathcal{P}_n, k = 1, ..., n$). If for some $M \in \mathcal{B}(\mathcal{H})$ and any $\epsilon > 0$, there is a partitioning $\mathcal{P}_n$ of $\mathcal{P}$, such that $\|M - M_n(\mathcal{P})\| < \epsilon$, then M is called *the right multiplicative integral along chain $\mathcal{P}$*. We write

$$M = \int_{\mathcal{P}}^{\rightarrow} (I + dF(P)).$$

For more details about the multiplicative integrals see [52, p. 493]. In particular, if ϕ is a scalar function defined and bounded on $\mathcal{P}$ and $B \in \mathcal{B}(\mathcal{H})$ then the integral

$$\int_{\mathcal{P}}^{\rightarrow} (I + \phi(P)B dP)$$

means the limit in the above pointed sense of the products

$$\overset{\rightarrow}{\prod_{1 \leq k \leq n}} (I + \phi(P_k)B\Delta P_k).$$

Lemma 9.11. *Let V be a compact quasi-nilpotent operator with a maximal invariant chain $\mathcal{P}$. Then*

$$(I - V)^{-1} = \int_{\mathcal{P}}^{\rightarrow} (I + V dP).$$

Proof. The sequence of the operators

$$V_n = \sum_{k=1}^{n} P_{k-1} V \Delta P_k \quad (P_k \in \mathcal{P}_n, k = 1, ..., n) \tag{9.1}$$

tends to V in the operator norm as $n \to \infty$, since

$$V - V_n = \sum_{k=1}^{n} \Delta P_k V \Delta P_k \to 0$$

in the operator norm thanks to the well known Lemma I.3.1 [55]. Besides, V_n are nilpotent, since

$$V_n^n = V_n^n P_n = V_n^{n-1} P_{n-1} V_n = V_n^{n-2} P_{n-2} V_n P_{n-1} V_n = \ldots$$

$$= V_n P_1 \cdots V_n P_{n-1} V_n = 0.$$

Due to Lemma 3.14,

$$(I - V_n)^{-1} = \overrightarrow{\prod_{2 \leq k \leq n}} (I + V_n \Delta P_k).$$

But $(I - V_n)^{-1} \to (I - V)^{-1}$ in the operator norm, cf. [18, p. 585, Lemma VII.6.3]. Hence the required result follows. Q. E. D.

Let the diagonal part of a $\mathcal{P}$-triangular operator A have the form $D = \int_{\mathcal{P}} \phi(P) dP$, where $\phi(P)$ is a scalar valued function of $P \in \mathcal{P}$. So

$$A = \int_{\mathcal{P}} \phi(P) dP + V \tag{9.2}$$

where V is the nilpotent part of A.

Theorem 9.9. *Let a $\mathcal{P}$-triangular operator A be represented by (9.2). Then*

$$(A - \lambda I)^{-1} = \int_{\mathcal{P}} \frac{dQ}{\phi(Q) - \lambda} \overrightarrow{\int_{\mathcal{P}}} \left(I + \frac{V dP}{\phi(P) - \lambda} \right) \quad (\lambda \notin \sigma(A)). \tag{9.3}$$

Proof. We have

$$(A - \lambda I)^{-1} = (D + V - \lambda I)^{-1} = (D - \lambda I)^{-1} (I + V(D - \lambda I)^{-1})^{-1} \quad (\lambda \notin \sigma(A)). \tag{9.4}$$

Due Corollary 8.1 $V(D - \lambda I)^{-1}$ is quasi-nilpotent. By the previous lemma

$$(I + V(D - \lambda I)^{-1})^{-1} = \overrightarrow{\int_{\mathcal{P}}} (I + V(D - \lambda I)^{-1} dP) \tag{9.5}$$

But

$$(D - \lambda I)^{-1} = \int_{\mathcal{P}} (\phi(P) - \lambda)^{-1} dP \qquad (9.6)$$

and therefore, $(D - \lambda I)^{-1} dP = (\phi(P) - \lambda)^{-1} dP$. Now (9.4) and (9.5) prove the theorem. Q. E. D.

Making use of Corollary 8.4 and Theorem 9.9 we have

Corollary 9.4. *Let $A \in \mathcal{B}(\mathcal{H})$, $A_I \in SN_p$ $(1 \le p < \infty)$ and $\sigma(A)$ be real. Then there are a maximal chain $\mathcal{P}$ and a nondecreasing scalar valued function $a(P)$ defined on $\mathcal{P}$, such that (9.3) holds with $\phi(P) = a(P)$.*

It is not hard to check that $V dP = PV dP$ and $PV^* dP = 0$. Therefore, $V dP = 2iPV_I dP$. If $\sigma(A)$ is real, then $V_I = A_I$. Now Corollary 9.4 implies.

Corollary 9.5. *Let $A \in \mathcal{B}(\mathcal{H})$, $A_I \in SN_p$ $(1 \le p < \infty)$ and $\sigma(A)$ be real. Then there are a maximal chain $\mathcal{P}$ and nondecreasing scalar valued function $a(P)$ defined on $\mathcal{P}$, such that*

$$(A - \lambda I)^{-1} = \int_{\mathcal{P}} \frac{dQ}{a(Q) - \lambda} \int_{\mathcal{P}}^{\rightarrow} \left(I + \frac{2iPA_I dP}{a(P) - \lambda} \right) \quad (\lambda \notin \sigma(A)). \qquad (9.7)$$

Let $A = A^*$. Then from (9.7) we have (9.6) with $\phi(P) = a(P)$. Thus, Corollary 9.5 is a generalization of the representation for the resolvents of selfadjoint operators.

9.10   Comments to Chapter 9

Theorems 9.1, 9.5 and 9.6 are taken from the book [28] but the gaps in the proofs are filled. For applications of the results from Chapter 9 to partial integral operators see [31]. The relevant results can be found in [26].

The multiplicative representation of the resolvent via the spectral measure appears without the proof in [23]. In [28] the very short proof of the main result from [23] is given.

Chapter 10

Regular Functions of a Bounded Non-selfadjoint Operator

In this chapter we establish the norm estimates for functions of a bounded operator A in a Hilbert space $\mathcal{H}$, assuming that the functions are regular on the convex hull of the spectrum of A. Besides, it is assumed that either $A - A^*$ is a Hilbert-Schmidt operator, or $AA^* - I$ is a nuclear one. Applications of the obtained estimates to the Sylvester equation are also discussed. In addition, we investigate the rotations of invariant subspaces of operators with compact Hermitian components.

10.1 Preliminary results

Below $A/\mathcal{H}_1$ means the restriction of A onto $\mathcal{H}_1 \subset \mathcal{H}$.

We begin with the following lemma.

Lemma 10.1. Let P_k $(k = 0, ..., n < \infty; \ n > 2)$ be a finite chain of orthogonal projections in $\mathcal{H}$:

$$0 = P_0\mathcal{H} \subset P_1\mathcal{H} \subset .... \subset P_n\mathcal{H} = \mathcal{H}.$$

Let $B \in \mathcal{B}(\mathcal{H})$ be defined by

$$B = \sum_{k=1}^{n} \phi_k \Delta P_k + W \quad (\Delta P_k = P_k - P_{k-1}), \tag{1.1}$$

where ϕ_k $(k = 1, ..., n)$ are complex numbers and W is a compact operator satisfying the relations

$$P_{k-1}WP_k = WP_k \quad (k = 1, ..., n). \tag{1.2}$$

Then there is a sequence Q_l $(l = 1, 2, ...)$ of finite dimensional orthogonal projections converging strongly to I, such that the operators $T_l = Q_l B Q_l$ have the property

$$\sigma(T_l/(Q_l\mathcal{H})) \subseteq \sigma(B) \quad (l = 1, 2, ...). \tag{1.3}$$

Moreover, the nilpotent parts of T_l converge to W in the operator norm.

Proof. Put

$$S = \sum_{k=1}^{n} \phi_k \Delta P_k.$$

Clearly, the spectrum of S consists of numbers ϕ_k $(k = 1, ..., n)$. Due to condition (1.2) $W^n = 0$. In addition, W and S have joint invariant subspaces. Since $\Delta P_k W \Delta P_k = 0$, we have $\sigma(S) = \sigma(B)$. Consequently, ϕ_k $(k = 1, ..., n)$ are eigenvalues of B.

Furthermore, let $(\Delta P_k f, g)$ $(f, g \in \Delta P_k \mathcal{H})$ be the scalar product in $\Delta P_k \mathcal{H}$. Here $(.,.)$ is the scalar product in $\mathcal{H}$.

Let $\{e_m^{(k)}\}_{m=1}^{\infty}$ be an orthogonal normal basis in $\Delta P_k \mathcal{H}$. That is, $(\Delta P_k e_m^{(k)}, e_j^{(k)}) = 0$ if $j \neq m$ and $(\Delta P_k e_m^{(k)}, e_m^{(k)}) = 1$. Put

$$\Delta Q_l^{(k)} = \sum_{m=1}^{l} (., e_m^{(k)}) e_m^{(k)} \quad \text{and} \quad Q_l^{(j)} = \sum_{k=1}^{j} \Delta Q_l^{(k)} \quad (j = 1, ..., n; \ l = 1, 2, ...).$$

Clearly, $\Delta Q_l^{(k)}$ strongly converge to ΔP_k, and $Q_l^{(j)}$ strongly converge to P_j as $l \to \infty$. So $Q_l := Q_l^{(n)}$ strongly converge to I as $l \to \infty$. In addition, Q_l is nl-dimensional,

$$\Delta Q_l^{(k)} \Delta P_k = \Delta P_k \Delta Q_l^{(k)} = \Delta Q_l^{(k)}, Q_l^{(j)} P_j = P_j Q_l^{(j)} = Q_l^{(j)}$$

$(j = 1, ..., n; \ l = 1, 2, ...)$. The nl-dimensional operators

$$S_l = \sum_{k=1}^{n} \phi_k \Delta Q_l^{(k)} = S Q_l$$

strongly converge to S as $l \to \infty$ and $\sigma(S_l) = \sigma(S_l / Q_l \mathcal{H}) = \{\phi_k\} \subseteq \sigma(B)$. Besides, the multiplicity of ϕ_k as the eigenvalue of S_l is finite, while the multiplicity of ϕ_k as the eigenvalue of S is infinite $(k = 1, ..., n)$.

Due to condition (1.2), we obtain

$$W = \sum_{j=1}^{n} \sum_{k=1}^{n} \Delta P_j W \Delta P_k = \sum_{k=2}^{n} P_{k-1} W \Delta P_k.$$

Introduce the operators

$$W_l := Q_l W Q_l = \sum_{k=2}^{n} Q_l^{(k-1)} W \Delta Q_l^{(k)}.$$

Since $Q_l \xrightarrow{s} I$ as $l \to \infty$, operators $W_l \xrightarrow{s} W$. But W is compact and therefore operators W_l converge to W in the operator norm. Take into account that

$$W_l Q_l^{(j)} = \sum_{k=1}^{j} Q_l^{(k-1)} W \Delta Q_l^{(k)} = Q_l^{(j-1)} \sum_{k=1}^{n} Q_l^{(k-1)} W \Delta Q_l^{(k)}.$$

Hence

$$W_l Q_l^{(j)} = Q_l^{(j-1)} W_l Q_l^{(j)} \ (j = 1, ..., n).$$

So $Q_l^{(j)}$ $(j = 1, ..., n)$ are invariant projections of W_l and, in addition, W_l is nilpotent. Put $T_l := S_l + W_l = Q_l B Q_l$. Since W_l and S_l have joint invariant subspaces, we obtain $\sigma(T_l) = \sigma(S_l)$ and $\sigma(S_l/Q_l\mathcal{H}) = \{\phi_k\} \subseteq \sigma(B)$. The lemma is proved. Q. E. D.

Let

$$A \in \mathcal{B}(\mathcal{H}) \text{ and } \Im A \in SN_p \ (1 \leq p < \infty). \tag{1.4}$$

Then due to Theorem 8.8 and Corollary 8.2 A is a $\mathcal{P}$-triangular operator:

$$A = D + V \ (\sigma(A) = \sigma(D)),$$

where $D = D_A$ and $V = V_A$ are the diagonal and nilpotent parts of A, respectively, and $V \in SN_p$. We need the following lemma.

Lemma 10.2. *Let condition (1.4) hold. Then there are a sequence B_n ($n = 1, 2, ...$) of finite dimensional operators strongly converging to A and a sequence $Z_n \overset{s}{\to} I$ of orthogonal projections, such that*

$$B_n = Z_n B_n = B_n Z_n \text{ and } \sigma(B_n/(Z_n\mathcal{H})) \subseteq \sigma(A). \tag{1.5}$$

Moreover, the nilpotent parts of B_n converge to the nilpotent part of A in the norm of SN_p.

Proof. As in Section 8.4, $\mathcal{E}$ denotes the invariant subspace of A corresponding to the nonreal spectrum of A, $Q_\mathcal{E}$ is the projection of $\mathcal{H}$ onto $\mathcal{E}$, $C = A Q_\mathcal{E}$, $M = (I - Q_\mathcal{E}) A (I - Q_\mathcal{E})$, $\mathcal{P}_C$ is the maximal invariant chain of C, $\mathcal{P}_M$ is the maximal invariant chain of M, $\mathcal{P}_A = \mathcal{P} = \mathcal{P}_M \oplus \mathcal{P}_C$ is the maximal invariant chain of A ordered as in Section 8.4. Due to Theorem 8.8 the diagonal part of A is defined by

$$D_A = \int_{\mathcal{P}_M} a(P) dP + \sum_{k=1}^\infty \lambda_k(A) \Delta \hat{P}_k,$$

where $a(P)$ is a real nondecreasing function defined on $\mathcal{P}_M$, $\lambda_k(A)$ ($k = 1, 2, ...$), are nonreal eigenvalues of A. $\Delta \hat{P}_k = \hat{P}_k - \hat{P}_{k-1}$, $\hat{P}_k \in \mathcal{P}_C$ are the orthogonal invariant projections corresponding to the nonreal eigenvalues of A.

Let $P_k^{(n)}(M)$ $(k = 0, ..., n)$ be a partitioning of $\mathcal{P}_M$. Since $\hat{P}_n \overset{s}{\to} Q_\mathcal{E}$, D is a strong limit of the operator sums

$$D_n = \sum_{k=1}^{n-1} \lambda_k(A) \Delta \hat{P}_k + \lambda_n(A)(Q_\mathcal{E} - \hat{P}_{n-1}) + \sum_{k=1}^n a(P_k^{(n)}(M)) \Delta P_k^{(n)}(M),$$

as $n \to \infty$. Define the projections $P_k^{(n)}(A)$ $(k = 1, ..., 2n)$ by

$$P_k^{(n)}(A) = \hat{P}_k \ (k < n), P_n^{(n)}(A) = Q_\varepsilon$$

and

$$P_k^{(n)}(A) = P_{k-n}^{(n)}(M) + Q_\varepsilon \ (n < k \le 2n). \tag{1.6}$$

Then we can write

$$D_n = \sum_{k=1}^{2n} c(P_k^{(n)}(A))\Delta P_k^{(n)}(A) \ \ (\Delta P_k^{(n)}(A) = P_k^{(n)}(A) - P_{k-1}^{(n)}(A)), \tag{1.7}$$

where $c(P_k^{(n)}(A)) = \lambda_k(A)$ for $k = 1, ..., n$, and $c(P_{n+j}^{(n)}(A)) = a(P_j^{(n)}(M))$ for $j = 1, ..., n$.

In addition, denote

$$V_n = \sum_{k=1}^{2n} P_{k-1}^{(n)}(A)V\Delta P_k^{(n)}(A),$$

where V is the nilpotent part of A. Put $A_n = D_n + V_n$. Then relations (1.1) and (1.2) hold with $2n$ instead of n, $B = A_n$, $P_k = P_k^{(n)}(A)$, $\phi_k = c(P_k^{(n)}(A))$ and $W = V_n$. Besides,

$$\sigma(A_n) = \sigma(D_n) = \{c(P_k^{(n)}(A))\}_{k=1}^{2n} \subseteq \sigma(A). \tag{1.8}$$

Since V is compact quasi-nilpotent, due to [55, Theorem III.4.1], and Lemma 8.3 ,$V_n \to V$ uniformly and therefore $A_n \overset{s}{\to} A$ as $n \to \infty$.

Due to Lemma 10.1, for each $n < \infty$ there are finite dimensional projections Q_{ln} $(l = 1, 2, ...)$ strongly converging to I as $l \to \infty$, such that the operators $T_{ln} := Q_{ln}A_nQ_{ln}$ have the properties

$$\sigma(T_{ln}/Q_{ln}\mathcal{H}) \subseteq \sigma(A_n) \subseteq \sigma(A) \ (l, n = 1, 2, ...).$$

Put $Z_n = Q_{nn}$, $B_n = T_{nn}$. Then $B_nZ_n = Z_nB_n$ and $\sigma(B_n/Z_n\mathcal{H}) \subseteq \sigma(A)$. Since $A_n \overset{s}{\to} A$, we have $B_n \overset{s}{\to} A$. Moreover, the nilpotent parts $\hat{V}_n$ of B_n uniformly converge to V. But $N_p(\Im V) \le N_p(\Im A) + N_p(\Im D))$ and due to Lemma 8.8, $V \in SN_p$. Thus $\hat{V}_n \to V$ in the norm $N_p(.)$ as $n \to \infty$. This finishes the proof. Q. E. D.

Let B_n be as in the previous lemma. Making use of (1.5), we have $\rho(B_n/(Z_n\mathcal{H}), \lambda) \ge \rho(A, \lambda)$ $(\lambda \notin \sigma(A))$. For any finite $p \ge 1$, there is an integer $\nu \ge 1$, such that $\Im A \in SN_p$ implies $\Im A \in SN_{2\nu}$. Then by Theorem 9.5,

$$\|(B_n - \lambda Z_n)^{-1}\| \le \sum_{m=0}^{\nu-1} \sum_{k=0}^{\infty} \frac{(c_\nu N_{2\nu}(\Im B_n))^{k\nu+m}}{\sqrt{k!}\rho^{k\nu+m+1}(A, \lambda)},$$

where the constant c_ν depends only on ν. Since $\Im B_n \overset{s}{\to} \Im A$, with $d_\nu(A) = c_\nu \sup_n N_{2\nu}(\Im B_n) < \infty$ we have

$$\|(B_n - \lambda Z_n)^{-1}\| \le \sum_{m=0}^{\nu-1} \sum_{k=0}^{\infty} \frac{d_\nu^{k\nu+m}(A)}{\sqrt{k!}\rho^{k\nu+m+1}(A,\lambda)}. \quad (1.9)$$

For any $s_0 \in \sigma(A)$ we have $|s_0 - \lambda| \ge \rho(A,\lambda)$. In addition B_n and $I - Z_n$ are mutually orthogonal. Hence,

$$\|(B_n + s_0(I - Z_n) - \lambda I)^{-1}\|^2 =$$
$$\|(B_n - \lambda Z_n + (s_0 - \lambda)(I - Z_n))^{-1}\|^2$$
$$\le \max\{\|(B_n - \lambda Z_n)^{-1}\|^2, |s_0 - \lambda|^{-2}\}.$$

Thus, from (1.9) we get

$$\|(B_n - \lambda Z_n + s_0(I - Z_n) - \lambda I)^{-1}\| \le \sum_{m=0}^{\nu-1} \sum_{k=0}^{\infty} \frac{d_\nu^{k\nu+m}(A)}{\sqrt{k!}\rho^{k\nu+m+1}(A,\lambda)}.$$

Since $I - Z_n \overset{s}{\to} 0$, due to Corollary 1.2 and Lemma 10.2, for any f regular on a simply connected set $M \supset \sigma(A)$, we obtain

$$f(B_n + s_0(I - Z_n)) \overset{s}{\to} f(A) \quad (s_0 \in \sigma(A)).$$

But

$$R_\lambda(B_n + s_0(I - Z_n)) = Z_n R_\lambda(B_n) + \frac{1}{s_0 - \lambda}(I - Z_n) \quad (\lambda \notin \sigma(A)) \quad (1.10)$$

and therefore,

$$f(B_n + s_0(I - Z_n)) = Z_n f(B_n) + f(s_0)(I - Z_n).$$

Thus

$$f(B_n + s_0(I - Z_n)) - Z_n f(B_n) = f(s_0)(I - Z_n) \overset{s}{\to} 0.$$

Similarly, due to Corollary 1.1 and Lemma 10.2,

$$R_\lambda(B_n + s_0(I - \tilde{Z}_n)) \overset{s}{\to} R_\lambda(A).$$

Taking into account (1.10), we have

$$R_\lambda(B_n + s_0(I - Z_n)) - Z_n R_\lambda(B_n) = \frac{1}{s_0 - \lambda}(I - Z_n) \to 0.$$

But

$$Z_n R_z(B_n) = R_z(B_n) Z_n.$$

So we arrive at

Corollary 10.1. *Let condition (1.4) hold. Then there are a sequence B_n $(n = 1, 2, ...)$ of finite dimensional operators strongly converging to A, and a sequence $Z_n \overset{s}{\to} I$ of orthogonal projections, such that (1.5) holds. Moreover,*

$$Z_n R_\lambda(B_n) = R_\lambda(B_n) Z_n \overset{s}{\to} R_\lambda(A) \quad (\lambda \notin \sigma(A)),$$

and for any f regular on a simply connected open set $M \supset \sigma(A)$, we have $Z_n f(B_n) = f(B_n) Z_n \overset{s}{\to} f(A)$.

10.2 Functions of an operator with a Hilbert-Schmidt component

In this section it is assumed that

$$A \in \mathcal{B}(\mathcal{H}) \text{ and } \Im A \in SN_2. \tag{2.1}$$

Recall that

$$g_I(A) = \sqrt{2}[N_2^2(\Im A) - \sum_{k=1}^{\infty}(\Im \lambda_k(A))^2]^{1/2}.$$

Theorem 10.1. *Let condition (2.1) hold and $f(z)$ be regular on a neighborhood of $co(A)$. Then*

$$\|f(A)\| \leq \sup_{\lambda \in \sigma(A)} |f(\lambda)| + \sum_{k=1}^{\infty} \sup_{\lambda \in co(A)} |f^{(k)}(\lambda)| \frac{g_I^k(A)}{(k!)^{3/2}}. \tag{2.2}$$

This theorem is sharp: if A is normal, then from Theorem 10.1 we have $\|f(A)\| = \sup_{\lambda \in \sigma(A)} |f(\lambda)|$. In this case it is only required that f is bounded on $\sigma(A)$.

Proof of Theorem 10.1: take operators B_n ($n = 1, 2, ...$) as in Lemma 10.2. Then $\sigma(B_n) \subseteq \sigma(A)$ and $co(B_n) \subseteq co(A)$. Making use of Theorem 3.5, we arrive at the inequality

$$\|Z_n f(B_n)\| \leq \sup_{\lambda \in \sigma(A)} |f(\lambda)| + \sum_{k=1}^{\infty} \sup_{\lambda \in co(A)} |f^{(k)}(\lambda)| \frac{g^k(B_n)}{(k!)^{3/2}}. \tag{2.3}$$

But due to Lemmas 3.1 and 3.2 we have

$$g(B_n) = N_2(\hat{V}_n) = g_I(B_n), \tag{2.4}$$

where $\hat{V}_n$ is the nilpotent part of B_n. We need the following result, which is due to (2.4), and Lemmas 10.2 and 9.3.

Corollary 10.2. *One has*

$$g_I(B_n) = N_2(\hat{V}_n) \to N_2(V) = g_I(A). \tag{2.5}$$

The assertion of Theorem 10.1 follows from inequality (2.3) and Corollary 10.2, since by Corollary 1.2 we have

$$\|f(A)\| \leq \sup_n \|f(B_n)\|.$$

Q. E. D.

In Section 10.4 below we suggest an additional proof of Theorem 10.1.

Remark 10.1. Let A_n $(n = 1, 2, \ldots)$ be an arbitary sequence of finite dimensional operators strongly converging to A. If f is regular on the disc $\{z \in \mathbf{C} : |z| \leq m_0\}$ with $m_0 > \sup_n \|A_n\|$, then the proof of (2.2) can be considerably simplified. Indeed, in this case (2.2) is valid due to Theorem 3.5 and Corollary 1.3.

10.3 Integral models of quasi-nilpotent operators

The aim of this section is to recall the well-known result which asserts that roughly speaking, each quasi-nilpotent Hilbert-Schmidt operator is unitarily equivalent to a Volterra integral operator. That result enables us to built an additional proof of Theorem 10.1.

An operator function $P(t)$, defined on a closed bounded set Ω of real numbers will be called a standard projection function, if the following conditions are fulfilled:

a) the values of $P(t)$ $(t \in \Omega)$ are orthogonal projections;

b) the projection function $P(t)$ is strictly increasing on Ω. I.e. $t_1 < t_2$ $(t_1, t_2 \in \Omega)$ implies $P(t_1) < P(t_2)$.

Let $\mathcal{P}$ be a closed chain, and $P(t)$ $(t \in \Omega)$ a standard projection function. We shall say that $P(t)$ is obtained by a *parametrization* of $\mathcal{P}$, if $\mathcal{P}$ coincides with the values of $P(t)$. A standard projection function $P(t)$ is said to be continuous if its values form a continuous chain. Let $\Omega = [a, b]$. Then $P(t)$ $(a \leq t \leq b)$ is said to be absolutely continuous if for every $f \in \mathcal{H}$, the (monotone) numerical function $(P(t)f, f)$ $(a \leq t \leq b)$ is absolutely continuous.

The following result holds [55, Theorem V.1.1].

Theorem 10.2. *Every closed chain $\mathcal{P}$ can be parametrized on an appropriate closed set $\Omega (\subseteq [0, 1])$. Under the parametrization the chain becomes a standard projection function $P(t)$ on Ω. If $\mathcal{P}$ is continuous, then its parametrization can be so chosen that $\Omega = [0, 1]$ and $P(t)$ $(0 \leq t \leq 1)$ is an absolutely continuous projection function.*

A system of vectors $\{d_j\}_{j=1}^r$ $(1 \leq r \leq \infty)$ is called a reproducing system for the chain $\mathcal{P}$, if the closed linear hull of the set of vectors Pd_j $(j = 1, 2, \ldots, r; P \in \mathcal{P})$ coincides with space $\mathcal{H}$. The smallest of the cardinalities of all possible reproducing systems for the chain is called the rank of this chain. Similarly the rank of $P(t)$ is defined.

We need the following concept. A linear operator $\tilde{A}$ acting in a Hilbert space $\tilde{\mathcal{H}} \supset \mathcal{H}$, will be called *an inessential extension* of operator $A \in \mathcal{B}(\mathcal{H})$ if $\tilde{A}h = Ah$ $(h \in \mathcal{H})$ and $\tilde{A}h = 0$ $(h \in \tilde{\mathcal{H}} \ominus \mathcal{H})$.

A standard projection function $P(t)$ $(t \in \Omega)$ is said to be an invariant function of an operator A if every subspace $P(t)\mathcal{H}$ is an invariant subspace of A, i.e. if $P(t)AP(t) = AP(t)$ $(t \in \Omega)$.

Theorem 10.3. *Let V be a quasi-nilpotent compact operator. If V has a maximal invariant chain of rank $r \leq \infty$, then some inessential extension of V has an absolutely continuous invariant projection-function of rank r.*

For the proof see [55, Theorem V.1.2].

Furthermore, let $L_r^2(0,1)$ be the Hilbert space of r-dimensional vector-functions $f(t)$ whose coordinates are scalar functions $f_k(t)$ $(k = 1, ..., r \leq \infty)$ defined on $[0,1]$ and the scalar product is defined by

$$(f,g)_{L_r^2(0,1)} = \int_0^1 \sum_{k=1}^r f_k(t)\overline{g}_k(t)dt \quad (f(t) = (f_k(t)), g(t) = (g_k(t))).$$

Let $\hat{P}(t)$ $(0 \leq t \leq 1)$ be the truncation projection function, defined by the conditions $\hat{P}(0) = 0, \hat{P}(1) = I$ and

$$(\hat{P}(s)f)(t) = \begin{cases} f(t) & \text{if } 0 \leq s < t, \\ 0 & \text{if } s < t \leq 1. \end{cases} \tag{3.1}$$

In the next fundamental (Sakhnovich's) theorem, cf. [55, Theorem V.2.2], $r \leq \infty$ denotes the rank of the maximal invariant chain of V.

Theorem 10.4. *Every quasi-nilpotent operator $V \in SN_2$ has an inessential extension, which is unitarily equivalent to a Volterra integral operator $\tilde{V}$ which acts in $L_r^2(0,1)$ according to the formula*

$$(\tilde{V}f)(t) = \int_t^1 K(t,s)f(s)ds \quad (f \in L_r^2(0,1)), \tag{3.2}$$

where $K(t,s) = (a_{jk}(t,s))_{j,k=1}^r$ $(0 \leq t \leq s \leq 1)$ is a matrix function, satisfying the condition

$$\sum_{j,k=1}^r \int_0^1 \int_t^1 |a_{jk}(t,s)|^2 ds\, dt < \infty.$$

Theorem 10.4 gives us an additional proof of the inequality

$$\|V^k\| \leq \frac{N_2^k(V)}{\sqrt{k!}} \quad \text{for any integer } k \geq 1 \tag{3.3}$$

(see Corollary 7.4).

Lemma 10.3. *Let $V \in \mathcal{B}(\mathcal{H})$ be a Hilbert-Schmidt quasi-nilpotent operator. Then inequality (3.3) is valid.*

Proof. Let $L^2(0,1)$ be the Hilbert space of scalar-valued functions defined on $[0,1]$ with the scalar product

$$(h,g) = \int_0^1 h(s)\overline{g}(s)ds.$$

Consider in $L^2(0,1)$ a Volterra Hilbert-Schmidt integral operator W defined by

$$(Wh)(t) = \int_t^1 K_1(t,s)h(s)ds, \; h \in L^2(0,1),$$

where $K_1(t,s)$ is a scalar kernel. It is well known that the kernel satisfies the condition

$$\int_0^1 \int_t^1 |K_1(t,s)|^2 ds \; dt = N_2^2(W) < \infty.$$

Employing Schwarz's inequality, we have

$$\|Wh\|^2 = \int_0^1 |\int_t^1 K_1(t,s)h(s)ds|^2 dt \le$$

$$\int_0^1 \int_t^1 |K_1(t,s)|^2 ds \int_t^1 |h(s_1)|^2 ds_1 dt.$$

Setting

$$w(t) = \int_t^1 |K_1(t,s)|^2 ds, \tag{3.4}$$

one can rewrite the latter inequality as

$$\|Wh\|^2 \le \int_0^1 w(t) \int_t^1 |h(s_1)|^2 ds_1 dt.$$

Using this inequality, we obtain

$$\|W^k h\|^2 \le \int_0^1 w(t) \int_t^1 |W^{k-1}h(s_1)|^2 ds_1 dt.$$

Once more apply Schwarz's inequality:

$$\|W^k h\|^2 \le \int_0^1 w(t) \int_t^1 w(s_1) \int_{s_1}^1 |W^{k-2}h(s_2)|^2 ds_2 \, ds_1 \, dt.$$

Continuing this process, we arrive at the inequality

$$\|W^k h\|^2 \le \int_0^1 w(t) \int_t^1 w(s_1) \int_{s_1}^1 w(s_2) \ldots$$

$$\int_{s_{k-2}}^{1} w(s_{k-1}) \int_{s_{k-1}}^{1} |h(s_k)|^2 ds_k \ldots ds_2 \, ds_1 \, dt.$$

Taking

$$\|h\|^2 = \int_0^1 |h(s)|^2 ds = 1,$$

we get

$$\|W^k\|^2 \le \int_0^1 w(t) \int_t^1 w(s_1) \ldots \int_{s_{k-2}}^1 w(s_{k-1}) ds_{k-1} \ldots ds_2 \, ds_1 \, dt. \quad (3.5)$$

It is simple to see that

$$\int_0^1 w(t) \int_t^1 w(s_1) \ldots \int_{s_{k-2}}^1 w(s_{k-1}) ds_{k-1} \ldots ds_2 \, ds_1 \, dt =$$

$$\int_0^1 \int_{z_1}^1 \ldots \int_{z_{k-1}}^1 dz_k \, dz_{k-1} \ldots dz_1 = \frac{m^k}{k!}.$$

Here

$$z_k = z_k(s_k) := \int_{s_k}^1 w(s_{k-1}) ds_{k-1}, \text{ and } m = \int_0^1 w(s) ds.$$

Thus (3.5) gives us

$$\|W^k\|^2 \le \frac{(\int_0^1 w(s) ds)^k}{k!} \quad (k = 1, 2, \ldots).$$

But according to (3.4) $m = N_2^2(W)$. Thus,

$$\|W^k\|^2 \le \frac{N_2^{2k}(W)}{k!} \quad (k = 1, 2, \ldots) \quad (3.6)$$

and for W the required inequality is proved.

Due to Theorem 10.4, there exists an inessential extension of the operator V which is unitarily equivalent to some Volterra Hilbert-Schmidt operator $\tilde{V}$ acting in the space $L_r^2(0, 1)$ according to (3.2).

Repeating our above arguments to the operator $\tilde{V}$, we obtain the following inequality which is similar to (3.6):

$$\|\tilde{V}^k\|^2 \le \frac{N_2^{2k}(\tilde{V})}{k!} \quad (k = 1, 2, \ldots).$$

Since $\|V^k\| = \|\tilde{V}^k\|$ and $N_2(V) = N_2(\tilde{V})$, we get the required inequality. Q. E. D.

10.4 Operators with Hilbert-Schmidt nilpotent parts

Assume that

$$A \in \mathcal{B}(\mathcal{H}), \Im A \in SN_p \ (p \geq 2). \tag{4.1}$$

Due to Theorem 8.8 A is $\mathcal{P}$-triangular and representations (2.3) and (2.4) hold. Our aim in this section is to prove the following theorem, in which we do not require that $\Im A \in SN_2$.

Theorem 10.5. *Let condition (4.1) hold and the nilpotent part V of A be a Hilbert-Schmidt operator. Let f be a function holomorphic on a neighborhood of $co(A)$. Then*

$$\|f(A)\| \leq \sup_{\lambda \in \sigma(A)} |f(\lambda)| + \sum_{k=1}^{\infty} \sup_{\lambda \in co(A)} |f^{(k)}(\lambda)| \frac{N_2^k(V)}{(k!)^{3/2}}. \tag{4.2}$$

To prove this theorem define D_n $(n = 1, 2, ...)$ as in (1.7). Then the operators $C_n = D_n + V = A - (D - D_n)$ strongly converge to A. As it was pointed in Section 10.2, $\sigma(A_n) = \sigma(D_n) \subseteq \sigma(A)$.

For simplicity put in (1.7)

$$c(P_k^{(n)}(A)) = c_k, P_k^{(n)}(A) = P_k, \Delta P_k^{(n)}(A) = \Delta P_k$$

and denote

$$I_{j_1 \dots j_{k+1}} = \frac{(-1)^{k+1}}{2\pi i} \int_L \frac{f(\lambda) d\lambda}{(c_{j_1} - \lambda) \dots (c_{j_{k+1}} - \lambda)},$$

where L is a Jordan contour surrounding $\sigma(A)$.

Lemma 10.4. *Let D_n be defined by (1.7), $C_n = D_n + V$ and $V \in SN_2$. Let f be holomorphic on a neighborhood of $co(C_n)$. Then*

$$\|f(C_n) - f(D_n)\| \leq \sum_{k=1}^{\infty} \frac{J_k N_2^k(V)}{\sqrt{k!}},$$

where $J_k = \max\{|I_{j_1 \dots j_{k+1}}| : 1 \leq j_1 < \dots < j_{k+1} \leq n\}$.

Proof. We have

$$R_\lambda(C_n) = (D_n + V - \lambda I)^{-1} = (I + R_\lambda(D_n)V)^{-1} R_\lambda(D_n)$$

$$= \sum_{k=0}^{\infty} (-1)^k (R_\lambda(D_n)V)^k R_\lambda(D_n),$$

since $R_\lambda(D_n)V$ is quasi-nilpotent due to Corollary 8.1. Thus

$$f(C_n) - f(D_n) = -\frac{1}{2\pi i}\int_L f(\lambda)(R_\lambda(A) - R_\lambda(D_n))d\lambda = \sum_{k=1}^{\infty} B_k, \quad (4.3)$$

where

$$B_k = (-1)^{k+1}\frac{1}{2\pi i}\int_L f(\lambda)(R_\lambda(D_n)V)^k R_\lambda(D_n)d\lambda.$$

But

$$R_\lambda(D_n) = \sum_{j=1}^{n}\frac{\Delta P_j}{c_j - \lambda}.$$

Besides,

$$\Delta P_j V \Delta P_k = 0 \quad (j \ge k).$$

Consequently,

$$B_k = \sum_{j_1=1}^{j_2-1}\Delta P_{j_1}V\sum_{j_2=1}^{j_3-1}\Delta P_{j_2}V\ldots\sum_{j_k=1}^{j_{k+1}-1}V\sum_{j_{k+1}=1}^{n}\Delta P_{j_{k+1}}I_{j_1 j_2\ldots j_{k+1}}.$$

Due to Theorem 10.4, there exists an inessential extension of V is unitarily equivalent to $\tilde{V}$ defined in $L_r^2(0,1)$ by (3.2). In addition, the projection-function $\hat{P}$ defined by (3.1) is invariant for $\hat{V}$. So $\hat{P}$ is unitarily equivalent to P, and B_k is unitarily equivalent to

$$\hat{B}_k := \sum_{j_1=1}^{j_2-1}\Delta\hat{P}_{j_1}\tilde{V}\sum_{j_2=1}^{j_3-1}\Delta\hat{P}_{j_2}\tilde{V}\ldots\sum_{j_k=1}^{j_{k+1}-1}\tilde{V}\sum_{j_{k+1}=1}^{n}\Delta\hat{P}_{j_{k+1}}I_{j_1 j_2\ldots j_{k+1}}$$

$$(\Delta\hat{P}_{j_{k+1}} = \hat{P}(t_k) - \hat{P}(t_{k-1}), 0 = \hat{P}(0) < \hat{P}(t_1) < \ldots < \hat{P}(t_n) = I).$$

Let $|K(t,s)|$ be the matrix whose entries $|a_{jk}(t,s)|$ $(j,k = 1,\ldots,r)$ are the absolute values of entries of matrix $K(t,s)$. Define $|\tilde{V}|$ by

$$|\tilde{V}|f(t) = \int_t^1 |K(t,s)|f(s)ds \quad (f \in L_r^2(0,1)).$$

Let $\mathcal{K}^+$ be the cone of vector functions in $L_r^2(0,1)$ with non-negative coordinates. Then $|\tilde{V}|$ and $\Delta\hat{P}_k$ are positive, i.e. they map $\mathcal{K}^+$ into itself. We have

$$|\tilde{B}_k f| \le J_k\sum_{j_1=1}^{j_2-1}\Delta\hat{P}_{j_1}|\tilde{V}|\sum_{j_2=1}^{j_3-1}\Delta\hat{P}_{j_2}|\tilde{V}|\ldots|\tilde{V}|\sum_{j_{k+1}=1}^{n}\Delta\hat{P}_{j_{k+1}}|f| = J_k|\tilde{V}|^k|f|.$$

Here $|f|$ is the vector function whose coordinates are the absolute values of $f \in L_r^2(0,1)$ and the inequality is understood in the sense of $\mathcal{K}^+$.

Thus $\|\hat{B}_k\| \le J_k \||\tilde{V}|^k\|$. According to Lemma 10.3,

$$\|\hat{B}^k\| \le \frac{N_2^k(|\tilde{V}|)J_k}{\sqrt{k!}}.$$

Since $\|\hat{B}_k\| = \|B_k\|$ and $N_2(V) = N_2(\tilde{V}) = N_2(|\tilde{V}|)$, due to (4.3) we get the required result. Q. E. D.

Proof of Theorem 10.5: Making use of Lemma 3.8, we get

$$J_k \le \sup_{\lambda \in co(C_n)} \frac{|f^{(k)}(\lambda)|}{k!}.$$

Since $\sigma(C_n) \subseteq \sigma(A)$, from the latter lemma it follows

$$\|f(C_n)\| \le \sup_{\lambda \in \sigma(A)} |f(\lambda)| + \sum_{k=1}^{\infty} \sup_{\lambda \in co(A)} |f^{(k)}(\lambda)| \frac{N_2^k(V)}{(k!)^{3/2}}.$$

Since $C_n \overset{s}{\to} A$, due to Theorem 9.5 and Corollary 1.2, $f(C_n) \overset{s}{\to} f(A)$ and $\|f(A)\| \le \sup_n \|f(C_n)\|$. This proves the theorem. Q. E. D.

Theorem 10.5 gives us *an additional proof of Theorem 10.1:* since the equality $N_2(V) = g_I(A)$ is valid, provided $\Im A \in SN_2$ (see Lemma 9.3), Theorem 10.1 from follows Theorem 10.5. Q. E. D.

10.5 Functions of an operator close to a unitary one

Let U_0 be a unitary operator and

$$T = U_0 + K, \text{ where } K \in SN_1. \tag{5.1}$$

In addition, assume that

$$T \text{ has a regular point on the unit circle } \{z \in \mathbf{C} : |z| = 1\}. \tag{5.2}$$

From (5.1) it follows

$$T^*T - I \in SN_1. \tag{5.3}$$

Again put

$$\vartheta(T) := [\text{trace } (T^*T - I) - \sum_{k=1}^{\infty}(|\lambda_k(T)|^2 - 1)]^{1/2}.$$

Theorem 10.6. *Let conditions (5.2) and (5.3) hold. Let $f(z)$ be regular on $co(T)$. Then*

$$\|f(T)\| \le \sup_{\lambda \in \sigma(T)} |f(\lambda)| + \sum_{k=1}^{\infty} \sup_{\lambda \in co(T)} |f^{(k)}(\lambda)| \frac{\vartheta^k(T)}{(k!)^{3/2}}. \tag{5.4}$$

Before proving this theorem note that it is sharp: if T is normal, then $\vartheta(T) = 0$ and $\|f(T)\| = \sup_{\lambda \in \sigma(T)} |f(\lambda)|$.

Proof of Theorem 10.6: Without loss of generality suppose that T has on the unit circle the regular point $\lambda_0 = 1$. If that point is e^{it} ($0 < t \le 2\pi$), then we can consider the operator $e^{-it}T$. So it is assumed that the operator

$$I - T \text{ is boundedly invertible.} \tag{5.5}$$

According to Lemma 8.9, if (5.3) and (5.5) holds, then the operator

$$A = i(I - T)^{-1}(I + T) \tag{5.6}$$

is bounded and satisfies the condition

$$A - A^* \in SN_1. \tag{5.7}$$

As Theorem 8.9 asserts, operators T and A are $\mathcal{P}$-triangular: $T = D_T + V_T$ and $A = D_A + V_A$, where D_T and D_A are the diagonal parts of T and A, respectively, V_T and V_A are the nilpotent parts of T and A, respectively. The transformation inverse to (5.6) can be defined by the formula

$$T = (A - iI)(A + iI)^{-1} = (D + V - iI)(D + V + iI)^{-1}. \tag{5.8}$$

Besides,

$$D_T = (D_A - iI)(D_A + iI)^{-1} \text{ and } V_T = T - D_T$$

(see the equality (5.7) in the proof of Theorem 8.9). We can write

$$D_T = (D_A - iI)(D_A + iI)^{-1} \text{ and } V_T = T - D_T.$$

Moreover, according to (2.4) and (2.5) D_A is a strong limit of the operators

$$D_{A_n} = D_n = \sum_{k=1}^{n} c(P_k^{(n)}(A)) \Delta P_k^{(n)}(A).$$

Put

$$D_{T_n} = \sum_{k=1}^{n} b_k \Delta P_k^{(n)}(A),$$

where

$$b_k = \frac{c(P_k^{(n)}(A)) - i}{c(P_k^{(n)}(A)) + i}.$$

We have $D_T = \hat{f}(D_A)$ and $D_{T_n} = \hat{f}(D_{A_n})$, where

$$\hat{f}(z) = \frac{z - i}{z + i}.$$

Besides $\sigma(A_n) \subseteq \sigma(A)$. So by the spectrum mapping theorem $\sigma(T_n) \subseteq \sigma(T)$. Since $D_{A_n} \to A$ strongly and $D_{T_n} = \hat{f}(D_{A_n})$, due to Corollary 1.2, taking into account that D_{A_n}, D_A are normal, we have

$$D_{T_n} = \hat{f}(D_{A_n}) \overset{s}{\to} \hat{f}(D_A) = D_T.$$

Hence, $T_n = D_{T_n} + V_T \overset{s}{\to} T$.

Furthermore, put

$$\hat{I}_{j_1 \ldots j_{k+1}} = \frac{(-1)^{k+1}}{2\pi i} \int_C \frac{f(\lambda) d\lambda}{(b_{j_1} - \lambda) \ldots (b_{j_{k+1}} - \lambda)}$$

and

$$\hat{J}_k := \max\{|\hat{I}_{j_1 \ldots j_{k+1}}| : 1 \le j_1 < \ldots < j_{k+1} \le n\}.$$

Repeating the arguments of Lemma 10.4, we get

$$\|f(T_n) - f(D_{T_n})\| \le \sum_{k=1}^{\infty} \frac{\hat{J}_k N_2^k(V_T)}{\sqrt{k!}}, \tag{5.9}$$

provided f is a function holomorphic on a neighborhood of $co(T_n)$. Since $\sigma(T_n) \subseteq \sigma(T)$, from (5.9) due to Lemma 3.8 it follows

$$\|f(T_n)\| \le \sup_{\lambda \in \sigma(T)} |f(\lambda)| + \sum_{k=1}^{\infty} \sup_{\lambda \in co(T)} |f^{(k)}(\lambda)| \frac{N_2^k(V_T)}{(k!)^{3/2}}. \tag{5.10}$$

But $T_n \to T$ strongly. By Corollary 1.2 and Theorem 9.6, $f(T_n) \overset{s}{\to} f(T)$, and $\|f(T)\| \le \sup_n \|f(T_n)\|$.

To finish the proof we need only to apply the equality $N_2(V_T) = \vartheta(T)$, cf. Lemma 9.6. Q. E. D.

10.6   Examples

Example 10.1. Let condition (2.1) hold. Then by Theorem 10.1,

$$\|e^{At}\| \leq e^{\alpha(A)t} \sum_{k=0}^{\infty} \frac{t^k g_I^k(A)}{(k!)^{3/2}} \quad (t \geq 0),$$

where $\alpha(A) = \sup Re \ \sigma(A)$.

Example 10.2. By Theorem 10.1, under condition (2.1) one has

$$\|A^m\| \leq \sum_{k=0}^{m} \frac{m! r_s^{m-k}(A) g_I^k(A)}{(m-k)!(k!)^{3/2}} \quad (m = 1, 2, ...).$$

Recall that $r_s(A)$ is the spectral radius of an operator A.

Example 10.3. Let $f(z)$ be the principal branch of $\ln z$.

Assume that $r_0 := \inf_{z \in co(A)} |z| > 0$. Then

$$\frac{1}{k!}|(\ln z)^{(k)}| = \frac{1}{k|z|^k} \leq \frac{1}{kr_0^k} \quad (k \geq 1; \ z \in co(A)).$$

Thus under (2.1), Theorem 10.1 yields

$$\| \ln A\| \leq \max_{z \in \sigma(A)} |\ln z| + \sum_{k=1}^{\infty} \frac{g_I^k(A)}{k\sqrt{k!}r_0^k}.$$

Example 10.4. Let $f(z)$ be the principal branch of $z^{1/2}$. Again assume that $r_0 = \inf_{z \in co(A)} |z| > 0$. Then

$$\left| \frac{d^k}{dz^k} z^{1/2} \right| = \frac{c_k}{|z|^{k-\frac{1}{2}}} \leq \frac{c_k}{r_0^{k-\frac{1}{2}}(A)} \quad (k \geq 1; \ z \in co(A)),$$

where

$$c_1 = \frac{1}{2}, c_2 = \frac{1}{4}, c_k = \frac{1}{2^k} 3 \cdot 5 \cdots (2k-3) \quad (k \geq 3).$$

Thus Theorem 10.1 implies

$$\|\sqrt{A}\| \leq \sqrt{r_s(A)} + \sum_{k=1}^{\infty} \frac{c_k g_I^k(A)}{(k!)^{3/2} r_0^{k-\frac{1}{2}}},$$

provided condition (2.1) holds.

10.7 The Sylvester equation with non-selfadjoint operators

Now let us consider the Sylvester equation
$$AX - XB = C \quad (A \in \mathcal{B}(\mathcal{H}), B \in \mathcal{B}(\mathcal{E}), C \in \mathcal{B}(\mathcal{E}, \mathcal{H})) \tag{7.1}$$
under the conditions
$$\Im A \in SN_2, \Im B \in SN_2 \tag{7.2}$$
and
$$\beta(B) = \inf \Re \sigma(B) > \alpha(A). \tag{7.3}$$
Put
$$\gamma_\infty(a, A, B) := \sum_{j,k=0}^\infty \frac{(k+j)! g_I^k(A) g_I^j(B)}{a^{k+j+1}(k!j!)^{3/2}}$$
for a constant $a > 0$.

Lemma 10.5. *Let conditions (7.2) and (7.3) hold. Then*
$$\int_0^\infty \|e^{At}\| \|e^{-Bt}\| dt \le \gamma_\infty(\eta, A, B),$$
where $\eta := \beta(B) - \alpha(A)$.

Proof. Due to Example 10.1 we have
$$\|e^{As}\| \le e^{\alpha(A)s} \sum_{k=0}^\infty \frac{s^k g_I^k(A)}{(k!)^{3/2}}$$
and
$$\|e^{-Bs}\| \le e^{-\beta(B)s} \sum_{j=0}^{\tilde{n}-1} \frac{s^j g_I^j(B)}{(j!)^{3/2}}.$$
for all $s \ge 0$. So
$$\|e^{As}\| \|e^{-Bs}\| \le e^{-\eta s} \sum_{j,k=0}^\infty \frac{s^{k+j} g_I^k(A) g_I^j(B)}{(k!j!)^{3/2}}.$$
But
$$\int_0^\infty s^{k+j} e^{-\eta s} ds = \frac{(k+j)!}{\eta^{k+j+1}}.$$
This proves the lemma. Q. E. D.

Theorem 10.7. *Let conditions (7.2) and (7.3) hold. Then the unique solution X to equation (7.1) is subject to the inequality*
$$\|X\| \le \|C\| \gamma_\infty(\eta, A, B).$$

This result is due to Theorem 2.4 and the previous lemma.

If A is normal, then $g_I(A) = 0$ and therefore,

$$\gamma_\infty(\eta, A, B) := \sum_{j=0}^{\infty} \frac{g_I^j(B)}{\eta^{j+1}(j!)^{1/2}}.$$

If both A and B are normal, then $\gamma(\eta, A, B) \le \frac{1}{\eta}$.

Similarly, making use of Theorem 10.6 one can consider equation (7.1) under conditions (5.1), (5.2). Moreover, the Rosenblum Theorem and estimates for resolvents from Chapter 9 enable us to investigate solutions of the Sylvester equation under the conditions $\Im A, \Im B \in SN_{2p}$ for an integer $p > 1$.

10.8 Perturbations of invariant subspaces

Let P and F be orthogonal projections in $\mathcal{H}$ and $E = I - P$. As shown in Section 5.7, $\|EF\|$ is a good measure of separation between the subspaces $P\mathcal{H}$ and $F\mathcal{H}$.

The aim of this section is to prove the following

Theorem 10.8. *Let P and F be orthogonal invariant projections of $A \in \mathcal{B}(\mathcal{H})$ and $B \in \mathcal{B}(\mathcal{H})$, respectively. Let $E = I - P$ and the conditions (7.2) and*

$$\beta(EA) := \inf \Re\sigma(EA/E\mathcal{H}) > \alpha(BF) := \sup \Re\sigma(BF/F\mathcal{H}) \tag{8.1}$$

hold. Then

$$\|EF\| \le \|A - B\|\gamma_\infty(\delta, A, B), \tag{8.2}$$

where $\delta := \beta(EA) - \alpha(BF)$.

Proof. We have

$$AP = PAP \text{ and } BF = FBF. \tag{8.3}$$

Put $Y = E(A - B)F$. Due to (8.3) we can write $EA = EAE$ and

$$Y = EAEF - EFBF = A_E Z - ZB_F,$$

where $Z = EF$, $A_E = EA$, $B_F = BF$. So Z is a solution of the equation

$$A_E Z - ZB_F = Y.$$

Due to Theorem 10.7, taking into account that $\|Y\| \le \|A - B\|$, we have

$$\|Y\| \le \|A - B\|\gamma_\infty(\delta, A_E, B_F). \tag{8.4}$$

According to Theorem 8.8 A is $\mathcal{P}$-triangular: $A = D + V$, where D and V are normal and compact quasi-nilpotent operators, respectively, having the joint invariant subspaces. According to Lemma 9.3 we have $g_I(A) = N_2(V)$. Since P is invariant for V we have $(EV)^2 = EVEV = EV^2$. Similarly, $(EV)^m = EV^m$. So the operator EV is quasi-nilpotent and it is the nilpotent part of EA. Therefore, $g_I(A_E) = N_2(EV) \leq N_2(V) = g_I(A)$. Similarly, $g_I(B_F) \leq g_I(B)$. So $\gamma_\infty(\delta, A_E, B_F) \leq \gamma_\infty(\delta, A, B)$. Since $Z = EF$, inequality (8.4) implies the required result. Q. E. D.

In particular, if A is normal, then $g_I(A) = 0$ and (8.2) takes the form

$$\|EF\| \leq \|A - B\| \sum_{j=0}^{\infty} \frac{g_I^j(B)}{\delta^{j+1}(j!)^{1/2}}.$$

If both A and B are normal, then $\|EF\| \leq \|A - B\|/\delta$.

10.9 Comments to Chapter 10

Theorem 10.2 refines Theorem 7.10.1 from [28]. Applications of such types theorems to parabolic integro-differential equations are discussed in [50].

Theorem 10.8 firstly has been proved in [43]. It is a generalization (in the case of bounded operators) of the well-known Davis-Kahan theorem, cf. [16] (see also [15]). For more details about perturbations of invariant subspaces see for instance [7] and [49].

Chapter 11

Functions of an Unbounded Operator

This chapter deals with some functions of an unbounded operator in a separable Hilbert space $\mathcal{H}$. In particular, we establish sharp norm estimates for the resolvent and function e^{At} ($t \geq 0$) of an unbounded operator A with a compact Hermitian component. In addition, a norm estimate for the resolvent of an unbounded operator inverse to a Schatten-von Neumann one is derived. By that estimate we investigate the Hirsch functions of A. Fraction powers and operator logarithm are the examples of the Hirsch functions.

11.1 Boundedly perturbed selfadjoint operators

Consider the operator

$$A = S + T, \tag{1.1}$$

acting in $\mathcal{H}$, where T is bounded, $S = S^*$ and

$$\alpha(S) = \sup \sigma(S) < \infty. \tag{1.2}$$

Besides, S can be unbounded. From the Hilbert identity for resolvents it follows that

$$\|R_z(A)\| \leq \frac{\|R_z(S)\|}{1 - \|T\|\|R_z(S)\|}$$

provided $\|T\|\|R_z(S)\| < 1$. Since,

$$\|R_z(S)\| \leq \frac{1}{\Re z - \alpha(S)} \quad (\Re z > \alpha(S)),$$

for an integer n we have

$$\|R_n(A)\| \leq \frac{1}{n - \alpha(S)} \left(1 - \frac{\|T\|}{n - \alpha(S)}\right)^{-1} = \frac{1}{n - \alpha(S) - \|T\|}, \tag{1.3}$$

provided $n > \|T\| + \alpha(S)$. Thus $\|R_n(A)\| \to 0$ as $n \to \infty$.

Put

$$A_n = -n(I + n(A - In)^{-1}).\tag{1.4}$$

So A_n is bounded and

$$A_n v_0 = -n(A - nI)^{-1} A v_0 \quad (v_0 \in Dom(A)),\tag{1.5}$$

where $Dom(A)$ denotes the domain of A. Due to (1.3),

$$\|n(A - nI)^{-1}h + h\| = \|(A - In)^{-1}Ah\| \to 0 \quad (h \in Dom(A))\tag{1.6}$$

as $n \to \infty$. In addition, for any $n_0 > \alpha(S) + \|T\|$,

$$\sup_{n \geq n_0} n\|R_n(A)\| \leq \sup_{n \geq n_0} \frac{n}{n - \alpha(S) - \|T\|} < \infty.\tag{1.7}$$

Since $n\|(A - In)^{-1}\|$ are uniformly bounded for sufficiently large n, due to Theorem 1.1, from (1.6) it follows

$$n(A - In)^{-1}h \to -h \ (h \in \mathcal{H}),$$

and from (1.5) with $h = Av_0$ it follows that $A_n v_0 \to A v_0$. We thus have proved

Lemma 11.1. *Let A and A_n $(n = 1, 2, ...)$ be defined by (1.1) and (1.4), respectively. Let condition (1.2) hold. Then A_n strongly converge to A on $Dom(A)$ as $n \to \infty$.*

Since $A_n = \phi_n(A)$, where

$$\phi_n(z) = -n(1 + n(z - n)^{-1}) = -n(z - n)^{-1}z,$$

and $\phi_n(z) \to z$ $(n \to \infty)$ for any finite z, by the Spectral mapping theorem we have

$$\lim_{n \to \infty} \sigma(A_n) = \sigma(A).\tag{1.8}$$

In addition, assume that for $\lambda \notin \sigma(A)$ and sufficiently large n_0, let

$$\|(A_n - I\lambda)^{-1}\| \leq F(1/\rho(A_n, \lambda))\tag{1.9}$$

for all finite $n \geq n_0$. Here $F(x)$ is a continuous increasing function of $x \geq 0$ with $F(0) = 0$, $F(\infty) = \infty$. Then

$$((A_n - I\lambda)^{-1} - (A - \lambda I)^{-1})h = (A_n - \lambda I)^{-1}(A - A_n)(A - \lambda I)^{-1}h$$

$$= (A_n - \lambda I)^{-1}(A - \lambda I)^{-1}(A - A_n)h \to 0 \quad (n \to \infty)$$

for all $h \in Dom(A)$. In view of (1.8) and (1.9) the norms of $(A_n - I\lambda)^{-1}$ are uniformly bounded, due to Theorem 1.1, we get

Lemma 11.2. *Let A and A_n $(n = 1, 2, ...)$ be defined by (1.1) and (1.4), respectively. Let conditions (1.2) and (1.9) hold. Then $(A_n - \lambda I)^{-1} \xrightarrow{s} (A - I\lambda)^{-1}$ and*

$$\|(A - I\lambda)^{-1}\| \leq \overline{\lim}_{n \to \infty} \|(A_n - I\lambda)^{-1}\| \quad (\lambda \notin \sigma(A)).$$

Furthermore, under condition (1.2) the function (semigroup) $T(t) = e^{At}$ ($t \geq 0$) is defined as a continuously differentiable solution of the equation $dT(t)/dt = AT(t)$ with $T(0) = I$ and images in $Dom(A)$. We need the identity

$$e^{At} - e^{A_n t} = \int_0^t e^{A(t-s)}(A - A_n)e^{A_n s}ds = \int_0^t e^{A(t-s)}e^{As}ds(A - A_n) \ (t \geq 0).$$
$$(1.10)$$

Put $x(t) = e^{At}h$, $x_n(t) = e^{A_n t}$ ($h \in Dom(A)$). From the equality

$$\dot{x}(t) := \frac{d}{dt}x(t) = Ax(t)$$

it follows

$$\frac{d}{dt}(x(t), x(t)) = (\dot{x}(t), x(t)) + (x(t), \dot{x}(t)) = (Ax(t), x(t)) + (x(t), Ax(t))$$

$$= ((A + A^*)x(t), x(t)) \leq 2\alpha(A_R)(x(t), x(t)) \ (A_R = (A + A^*)/2).$$

Thus,

$$\|e^{At}\| \leq e^{\alpha(A_R)t} \ (t \geq 0). \tag{1.11}$$

Besides, $\alpha(A_R) \leq \alpha(S) + \|T\|$. Since $A_n h \to Ah$, $A_n^* h \to A^* h$ ($h \in Dom(A) = Dom(S)$), we have

$$((A_n + A_n^*)h, h) \to ((A + A^*)h, h) \leq 2\alpha(A_R)(h, h).$$

Since A and A_n commute, from (1.10) and (1.11) we obtain

$$\|(e^{At} - e^{A_n t})h\| \leq \int_0^t e^{\alpha(A_R)(t-s)}e^{c_0 s}ds\|(A - A_n)h\| \to 0$$

($n \to \infty$; $h \in Dom\ A$). Hence by virtue of Theorem 1.1 we get

Lemma 11.3. *Let A and A_n ($n = 1, 2, ...$) be defined by (1.1) and (1.4), respectively. Let condition (1.2) hold. Then*

$$e^{A_n t} \xrightarrow{s} e^{At} \ as \ n \to \infty \ (t > 0). \tag{1.12}$$

11.2 Unbounded operators with compact components

In this section it is assumed that A is a closed operator, satisfying the conditions

$$Dom(A^*) = Dom(A) \ \text{and} \ \sup \sigma(A_R) < \infty. \tag{2.1}$$

Recall that $A_I = \Im A = (A - A^*)/2i$.

It is also supposed that

$$A_I \in SN_{2p} \tag{2.2}$$

for an integer $p \geq 1$.

Let A_n be defined by (1.4). It is clear that

$$A_n - A_n^* = -n^2((A - In)^{-1} - (A^* - In)^{-1}) =$$

$$n^2(A - In)^{-1}(A - A^*)(A^* - In)^{-1}.$$

Hence according to (1.7) and (2.2), for sufficiently large n,

$$N_{2p}(A_n - A_n^*) \leq n^2\|(A - In)^{-1}\|^2 N_{2p}(A - A^*) < \infty.$$

In addition, since operators A_n strongly converge to A on $Dom(A)$, under (2.2), we easily have

$$N_{2p}(A_n - A_n^*) \to N_{2p}(A - A^*) \quad (n \to \infty). \tag{2.3}$$

Moreover, by (1.8),

$$\lim_{n \to \infty} \rho(A_n, \lambda) = \rho(A, \lambda).$$

Recall that $\tau_p(A)$ is defined in Section 9.5 and

$$\tau_p(A) \leq 2(1 + 2p)N_{2p}(A_I).$$

Now Theorem 9.5 and Lemma 11.2 imply

Theorem 11.1. *Let conditions (2.1) and (2.2) hold. Then for all $\lambda \notin \sigma(A)$ we have*

$$\|R_\lambda(A)\| \leq \sum_{m=0}^{p-1} \sum_{k=0}^{\infty} \frac{\tau_p^{pk+m}(A)}{\rho^{pk+m+1}(A, \lambda)\sqrt{k!}} \tag{2.4}$$

and

$$\|R_\lambda(A)\| \leq \sqrt{e} \sum_{m=0}^{p-1} \frac{\tau_p^m(A)}{\rho^{m+1}(A, \lambda)} \exp\left[\frac{\tau_p^{2p}(A)}{2\rho^{2p}(A, \lambda)}\right]. \tag{2.5}$$

Remark 11.1. If in Theorem 11.1 $p = 1$, then according to Theorem 9.1, one can take $\tau_1(A) = \sqrt{2}N_2(A_I)$, since $g_I^2(A_n) \leq N_2^2(A_n - A_n^*)/2$. So in the case $A_I \in SN_2$, for all $\lambda \notin \sigma(A)$ we have

$$\|R_\lambda(A)\| \leq \sum_{k=0}^{\infty} \frac{(\sqrt{2}N_2(A_I))^k}{\rho^{k+1}(A,\lambda)\sqrt{k!}} \tag{2.6}$$

and

$$\|R_\lambda(A)\| \leq \sqrt{e}\exp\left[\frac{N_2^2(A_I)}{\rho^2(A,\lambda)}\right]. \tag{2.7}$$

Furthermore, since A_R is selfadjoint, we can write $sv_{A_R}(A) \leq \|A_I\|$ and therefore $\alpha(A) < \infty$. Now from Lemma 11.3 and Example 10.1 it follows

Corollary 11.1. *Let condition (2.1) hold and $A - A^* \in SN_2$. Then $\alpha(A) < \infty$ and*

$$\|e^{At}\| \leq e^{\alpha(A)t} \sum_{k=0}^{\infty} \frac{t^k(\sqrt{2}N_2(A_I))^k}{(k!)^{3/2}} \quad (t \geq 0).$$

11.3 Resolvents of operators inverse to Schatten-von Neumann ones

This section deals with linear operators in a Hilbert space, whose inverse ones belong to the Schatten-von Neumann ideal of compact operators, and whose imaginary Hermitian components are bounded. A sharp norm estimate for the resolvents of the considered operators is derived. That estimate enables us to investigate spectrum perturbations and to establish bounds for the norms of the corresponding semigroups and Hirsch operator functions.

Throughout this section it is assumed that A is an invertible operator in $\mathcal{H}$ with the dense domain,

$$A^{-1} \in SN_p \text{ for an integer } p \geq 1, \tag{3.1}$$

$$\|A_I\| < \infty \text{ and } \beta(A_R) = \inf \sigma(A_R) > -\infty. \tag{3.2}$$

Note that instead of condition (3.1) in our reasonings below, we can require the condition $(A - aI)^{-1} \in SN_p$ for a regular a of A.

Introduce the notations

$$\theta_p(A) := 4N_p(A^{-1})\|A_I\|, \quad \rho(A,\lambda) := \inf_{s \in \sigma(A)} |\lambda - s|$$

and

$$\psi(A,\lambda) := \inf_{s \in \sigma(A)} \left|1 - \frac{\lambda}{s}\right| \quad (\lambda \in \mathbf{C}).$$

Theorem 11.2. *Let A be an invertible operator in $\mathcal{H}$ with the dense domain $Dom(A) = Dom(A^*)$, and conditions (3.1) and (3.2) hold. Then*

$$\|(A - \lambda I)^{-1}\| \leq \frac{1}{\rho(A, \lambda)} \Phi_p\left(\frac{\theta_p(A)}{\psi(A, \lambda)}\right) \quad (\lambda \notin \sigma(A)), \qquad (3.3)$$

where

$$\Phi_p(x) = \sum_{m=0}^{p-1} \sum_{k=0}^{\infty} \frac{x^{pk+m}}{\sqrt{k!}} \quad (x \geq 0).$$

The proof of this theorem is presented in the next section. If $A_I = 0$, then $\theta_p(A) = 0$ and we obtain the equality $\|(A - \lambda I)^{-1}\| = \rho^{-1}(A, \lambda)$, since $\Phi_p(0) = 0^0 = 1$.

11.4 Proof of Theorem 11.2

Lemma 11.4. *Under the hypothesis of Theorem 11.2 operator A^{-1} has a complete system of the roots vectors.*

Proof. Let c be a real constant satisfying $c \notin \sigma(A) \cup \sigma(A_R)$. Since A_R is selfadjoint, $sv_{A_R}(A) \leq \|A_I\|$ and therefore $\beta(A) \geq \beta(A_R) - \|A_I\| > -\infty$. So, one can take any $c < \beta(A_R) - \|A_I\|$. Obviously,

$$(A - cI)^{-1} = (I + i(A_R - cI)^{-1}A_I)^{-1}(A_R - cI)^{-1}. \qquad (4.1)$$

But $(A - cI)^{-1} = A^{-1}(I - cA^{-1})^{-1} \in SN_p$. So

$$(A_R - cI)^{-1} = (A - cI)^{-1}(I + i(A_R - cI)^{-1}A_I) \in SN_p.$$

Recall the Keldysh theorem, cf. [54, Theorem V. 8.1].

Theorem 11.3. *Let $B = H(I + S)$, where $H = H^* \in SN_p$ ($1 \leq p < \infty$) and S is a compact operator. In addition, let $Bh = 0$ only if $h = 0$. Then operator B has a complete system of the roots vectors in $\mathcal{H}$.*

By that theorem and (4.1) operator $(A - cI)^{-1}$ has a complete system of the roots vectors.

Since $(A - cI)^{-1}$ and A^{-1} commute, A^{-1} has a complete system of the roots vectors. As claimed. Q. E. D.

From the previous lemma it follows that there is the orthogonal normal (Schur) basis $\{e_k\}$, in which A^{-1} is represented by a triangular matrix (see [54, Lemma I.4.1]). Denote

$$P_k = \sum_{j=1}^{k} (., e_j)e_j.$$

Then

$$A^{-1}P_k = P_k A^{-1} P_k \quad (k = 1, 2, ...).$$

Besides,

$$\Delta P_k A^{-1} \Delta P_k = \lambda_k^{-1} \Delta P_k \quad (\lambda_k = \lambda_k(A) \in \sigma(A); \; k = 1, 2, ...; P_0 = 0).$$
$$(4.2)$$

Here $\Delta P_k = P_k - P_{k-1}$ and $\lambda_k = \lambda_k(A)$ are the eigenvalues of A taken with their multiplicities. Put

$$D = \sum_{k=1}^{\infty} \lambda_k \Delta P_k \text{ and } V = A - D.$$

Lemma 11.5. *Under conditions (3.1) and (3.2) one has* $\|V_I\| \leq 2\|A_I\|$
$(V_I = (V - V^*)/2i)$.

Proof. We have

$$A P_k f = P_k A P_k f \quad (k = 1, 2, ...; \; f \in Dom(A)). \qquad (4.3)$$

Indeed, $A^{-1} P_k$ is an invertible $k \times k$-matrix. Since $\Delta P_j P_k = 0$ for $j > k$, we have $0 = \Delta P_j A A^{-1} P_k = \Delta P_j A P_k A^{-1} P_k$. Hence $\Delta P_j A f = 0$ for any $f \in P_k \mathcal{H}$. This implies (4.3).

Due to (4.2) we can write

$$\Delta P_k = \Delta P_k \Delta P_k = \Delta P_k A A^{-1} \Delta P_k = \lambda_k^{-1} \Delta P_k A \Delta P_k.$$

Thus, $\Delta P_k A \Delta P_k = \Delta P_k D \Delta P_k = \lambda_k \Delta P_k$. But $\Delta P_k A P_{k-1} = \Delta P_k D P_{k-1} = 0$. So $\Delta P_k A P_k = \Delta P_k D P_k$ and thus $\Delta P_k V P_k = 0$. Hence,

$$V P_k = P_{k-1} V P_k. \qquad (4.4)$$

Put $A_n = A P_n, V_n = V P_n, D_n = D P_n$. Denote by A_{nI}, V_{nI} and D_{nI} the imaginary Hermitian components of A_n, V_n and D_n, respectively. We have $A_{nI} = V_{nI} + D_{nI}$ and $\|A_{nI}\| \leq \|A_I\|$. Moreover, due to the Weyl inequalities for bounded operators with compact imaginary components (see Lemma 8.7), we have $\|D_{nI}\| \leq \|A_{nI}\|$. Thus $\|V_{nI}\| = \|A_{nI} - D_{nI}\| \leq 2\|A_{nI}\|$. Hence, letting $n \to \infty$, we prove the lemma. Q. E. D.

Since $A^{-1} \in S_p$, by the Weyl inequalities for compact operators (see Lemma 7.3), we obtain

$$N_p^p(D^{-1}) = \sum_{k=1}^{\infty} \frac{1}{|\lambda_k|^p} \leq N_p^p(A^{-1}). \qquad (4.5)$$

In addition,

$$N_{2p}^{2p}(D^{-1/2}) = \sum_{k=1}^{\infty} \frac{1}{|\lambda_k|^p} = N_p^p(D^{-1}).$$

Thus

$$N_{2p}(VD^{-1}) = N_{2p}(VD^{-1/2}D^{-1/2}) \le \|VD^{-1/2}\| N_{2p}(D^{-1/2})$$

$$= \|VD^{-1/2}\| \sqrt{N_p(D^{-1})}. \tag{4.6}$$

But

$$\|VD^{-1/2}\|^{2p} \le \sum_{k=1}^{\infty} \|VD^{-1/2}e_k\|^{2p} = \sum_{k=1}^{\infty} \frac{1}{|\lambda_k|^p} \|Ve_k\|^{2p}. \tag{4.7}$$

Since $VP_k = P_{k-1}VP_k$, we have $P_kV^* = P_kV^*P_{k-1}$. But $P_{k-1}e_k = 0$. Thus, $P_{k-1}V^*e_k = P_kV^*P_{k-1}e_k = 0$. Consequently,

$$\|Ve_k\| = \|P_{k-1}Ve_k\| = \|P_{k-1}(V - V^*)e_k\| = 2\|P_{k-1}V_Ie_k\| \le 2\|V_Ie_k\|$$

and due to (4.7),

$$\|VD^{-1/2}\|^{2p} \le \sum_{k=1}^{\infty} \frac{1}{|\lambda_k|^p} \|2V_Ie_k\|^{2p} \le \|2V_I\|^{2p} \sum_{k=1}^{\infty} \frac{1}{|\lambda_k|^p}$$

$$= \|2V_I\|^{2p} N_p^p(D^{-1}).$$

Therefore

$$\|VD^{-1/2}\| \le 2\|V_I\| \sqrt{N_p(D^{-1})}.$$

Now due to (4.6) and Lemma 11.5 we arrive at our next result.

Lemma 11.6. *Under the hypothesis of Theorem 11.2 one has*

$$N_{2p}(VD^{-1}) \le 4N_p(D^{-1})\|A_I\|.$$

Now (4.6) implies

$$N_{2p}(VD^{-1}) \le \theta_p(A). \tag{4.8}$$

Lemma 11.7. *Under the hypothesis of Theorem 11.2, one has*

$$\|(A - \lambda I)^{-1}\| \le \frac{1}{\rho(A,\lambda)} \Phi_p \left(\frac{N_{2p}(VD^{-1})}{\psi(A,\lambda)} \right) \quad (\lambda \notin \sigma(A)).$$

Proof. For any $z \notin \sigma(A)$, it can be written

$$(A - Iz)^{-1} = (D + V - Iz)^{-1} = (D - Iz)^{-1}(I + VD^{-1}(I - D^{-1}z)^{-1})^{-1}$$

$$= (D - Iz)^{-1}(I + J(z))^{-1}, \tag{4.9}$$

where $J(z) := VD^{-1}(I - D^{-1}z)^{-1}$ $(z \notin \sigma(A))$. Due to the Lemma 11.6 $VD^{-1} \in SN_{2p}$ and due to (4.4) $J(z)P_k = P_{k-1}J(z)P_k$. So $J(z) \in SN_{2p}$ and it is a quasi-nilpotent operator, since $J(z)$ is a limit in the operator norm of nilpotent operators $J(z)P_k$ as $k \to \infty$ (see Lemma 8.1).

From Lemma 7.5 for a quasi-nilpotent operator $W \in SN_{2p}$, it follows

$$\|(I - W)^{-1}\| \leq \Phi_p(N_{2p}(W)).$$

Consequently,

$$\|(I + J(z))^{-1}\| \leq \Phi_p(N_{2p}(J(z))).$$

Since D is normal,

$$\|(D - z)^{-1}\| = \frac{1}{\rho(A, z)}$$

and

$$\|(I - D^{-1}z)^{-1}\| = \frac{1}{\rho(1, A^{-1}z)}.$$

But $\rho(1, A^{-1}z) = \psi(A, z)$. Thus

$$N_{2p}(J(z)) = N_{2p}(VD^{-1}(I - D^{-1}z)^{-1}) \leq N_{2p}(VD^{-1})\|(I - D^{-1}z)^{-1}\|$$

$$= \frac{N_{2p}(VD^{-1})}{\psi(A, z)}.$$

Now (4.9) proves the lemma. Q. E. D.

The assertion of Theorem 11.2 follows from (4.8) and Lemma 11.7. Q. E. D.

11.5   Hirsch type operator functions

Let the condition

$$\beta(A) = \inf\ Re\ \sigma(A) > 0 \qquad (5.1)$$

hold. Then

$$\rho(A, -t) = \inf_{s \in \sigma(A)} |s + t| \geq t + \beta(A) > 0 \quad (t \geq 0).$$

In addition, with $s = x + iy$, we have

$$\left| 1 + \frac{t}{s} \right|^2 = \left| 1 + \frac{t\bar{s}}{|s|^2} \right|^2 = \left| 1 + \frac{t(x - iy)}{|s|^2} \right|^2 =$$

$$(1 + tx|s|^{-2})^2 + (yt)^2|s|^{-4} \geq 1 \quad (t \geq 0; x \geq \beta(A)).$$

Thus

$$\psi(A, -t) = \inf_{s \in \sigma(A)} |1 + t/s| \geq 1.$$

Consequently, under conditions (5.1), (3.1) and (3.2), by Theorem 11.2,

$$\|(A + tI)^{-1}\| \leq \frac{\Phi_p(\theta_p(A))}{t + \beta(A)} \quad (t \geq 0). \qquad (5.2)$$

Define the function

$$f_H(A) = \int_0^\infty f(t)(A + tI)^{-1}dt, \qquad (5.3)$$

where f is a scalar function satisfying the condition

$$\int_0^\infty |f(t)|(1 + t)^{-1}dt < \infty. \qquad (5.4)$$

The formula (5.3) is a particular case of the Hirsch calculus, cf. [69]. We thus have proved the following

Theorem 11.4. *Let conditions (5.1), (3.1), (3.2) and (5.4) hold. Then*

$$\|f_H(A)\| \leq \Phi_p(\theta_p(A)) \int_0^\infty \frac{|f(t)|}{t + \beta(A)}dt.$$

Recall that the fractional power of A can be defined by the formula

$$A^{-\alpha} = \frac{sin\ (\pi\alpha)}{\pi} \int_0^\infty t^{-\alpha}(A + It)^{-1}dt \quad (0 < \alpha < 1), \qquad (5.5)$$

provided (5.1) holds, cf. [62, Section I.5.2, formula (5.8)]. Under conditions (3.1), (3.2) and (5.1) we have

$$\|A^{-\alpha}\| \leq \frac{sin\ (\pi\alpha)\ \Phi_p(\theta_p(A))}{\pi} \int_0^\infty \frac{dt}{t^\alpha(t + \beta(A))} \quad (0 < \alpha < 1). \qquad (5.6)$$

Furthermore, the operator logarithm can be represented as

$$\ln(A) = (A - I) \int_0^\infty (tI + A)^{-1} \frac{dt}{1+t},$$

provided the integral converges, cf. [69, Theorem 10.1.3]. Under conditions (3.1), (3.2) and (5.1) we have

$$\| \ln(A)x \| \le \Phi_p(\theta_p(A)) \int_0^\infty \frac{dt}{(t + \beta(A))(1+t)} \| (A - I)x \| \quad (x \in Dom(A)).$$
$$(5.7)$$

Now we are going to estimate the integral of the semigroup e^{-At} generated by $-A$. As it is well-known, cf. [62, Theorem I.1.3] the function $e^{-At}x_0$ with $x_0 \in Dom(A)$ is the inverse Laplace transform of $(A + zI)^{-1}x_0$. Due to the Parseval-Plancherel theorem [2, Theorem I.8.2, p. 45] we can write

$$\int_0^\infty \| e^{-At}x_0 \|^2 dt = \frac{1}{2\pi} \int_{-\infty}^\infty \| (A + iwI)^{-1}x_0 \|^2 dw. \qquad (5.8)$$

Note that

$$\rho^2(A, -iw) = \inf_{s=x+iy \in \sigma(A)} |s + iw|^2 \ge \inf_{x \ge \beta(A), |y| \le \|A_I\|} x^2 + (|y| - |w|)^2$$

$$\ge \beta^2(A) + (|w| - \|A_I\|)^2 \quad (w \in \mathbf{R}; |w| \ge \|A_I\|).$$

Hence $\rho(A, -iw) \ge \xi(w)$, where

$$\xi(w) = \begin{cases} \sqrt{\beta^2(A) + (|w| - \|A_I\|)^2} & \text{if } |w| \ge \|A_I\|, \\ \beta(A) & \text{if } |w| \le \|A_I\| \end{cases}.$$

Recall that

$$\psi(A, -iw) = \inf_{s \in \sigma(A)} \left| 1 + \frac{iw}{s} \right|.$$

With $s = x + iy \in \sigma(A)$ we have

$$\left| 1 + \frac{iw}{s} \right|^2 = \frac{|s + iw|^2}{|s|^2} = \frac{|x + i(y + w)|^2}{|s|^2} = \frac{x^2 + (y + w)^2}{x^2 + y^2} \ge$$

$$\inf_{x \ge \beta(A); |y| \le \|A_I\|} \frac{x^2}{x^2 + y^2} \ge \vartheta^2 \quad (w \in \mathbf{R}),$$

where

$$\vartheta : \frac{\beta(A)}{\sqrt{\beta^2(A) + \|A_I\|^2}}.$$

In view of Theorem 11,2,

$$\|(A + iwI)^{-1}\| \le \frac{\Phi_p(\theta_p(A)/\vartheta)}{\xi(w)} \quad (w \in \mathbf{R}). \tag{5.9}$$

Now (5.8) implies

$$\int_0^\infty \|e^{-At}x_0\|^2 dt \le \tilde{l}_p(A)\|x_0\|^2 \quad (x_0 \in Dom(A)), \tag{5.10}$$

where

$$\tilde{l}_p(A) := \frac{\Phi_p^2(\theta_p(A)/\vartheta)}{\pi} \int_0^\infty \frac{dt}{\xi^2(t)}.$$

Extending this inequality to all $x_0 \in \mathcal{H}$, we get $\int_0^\infty \|e^{-At}\|^2 dt < \infty$. Due to the well-known Theorem 4.1 from [72, p. 116] this implies the exponential stability of e^{-At}. We thus arrive at our next result.

Theorem 11.5. *Let conditions (3.1), (3.2) and (5.1) hold. Then e^{-At} is exponentially stable and satisfies inequality (5.10).*

Simple calculations show that

$$\int_0^\infty \frac{dt}{\xi^2(t)} = \frac{\|A_I\|}{\beta^2(A)} + \frac{\pi}{2\beta(A)}.$$

We thus have

$$\int_0^\infty \|e^{-At}x_0\|^2 dt \le \Phi_p^2(\theta_p(A)/\vartheta) \left(\frac{\|A_I\|}{\pi\beta^2(A)} + \frac{1}{2\beta(A)} \right) \|x_0\|^2. \tag{5.11}$$

11.6   Comments to Chapter 11

The results in Section 11.2 are particularly taken from [28]. Theorem 11.2 appears in [39]. The material in Section 11.5 is probably new.

Chapter 12

Similarity Condition Numbers of Unbounded Diagonalizable Operators

As in the finite dimensional case an operator A in a separable Hilbert space $\mathcal{H}$ is said to be a diagonalizable operator, if there are a boundedly invertible operator $T \in \mathcal{B}(\mathcal{H})$ and a normal operator D acting in $\mathcal{H}$, such that

$$THx = DTx \quad (x \in Dom(H)). \tag{0.1}$$

Besides, $\kappa_T = \|T^{-1}\|\|T\|$ is the *condition number*.

In this chapter we consider some unbounded diagonalizable operators and derive bounds for the condition numbers of the considered operators. We also discuss applications of the obtained bounds to spectrum perturbations and operator functions.

12.1 Condition numbers of operators with Schatten-von Neumann Hermitian components

Everywhere in this section H is an invertible operator in $\mathcal{H}$ with the following properties: $Dom(H) = Dom(H^*)$, and for some $r \in [1, \infty)$ and an integer $p \geq 1$, the conditions

$$H^{-1} \in SN_r \text{ and } H_I = (H - H^*)/2i \in SN_{2p} \tag{1.1}$$

hold. Numerous integro-differential operators satisfy these conditions.

Note that instead of the condition $H^{-1} \in SN_r$, in our reasonings below, one can require the condition $(H - aI)^{-1} \in SN_r$ for some point $a \notin \sigma(H)$.

Since H^{-1} is compact, $\sigma(H)$ is purely discrete.

It is assumed that *all the eigenvalues* $\lambda_j(H)$ *of* H *are different*. For a fixed integer m put

$$\delta_m(H) = \inf_{j=1,2,\dots;\, j \neq m} |\lambda_j(H) - \lambda_m(H)|.$$

It is further supposed that

$$\zeta_q(H) := \left[\sum_{j=1}^{\infty} \frac{1}{\delta_j^q(H)} \right]^{1/q} < \infty \qquad (1.2)$$

where

$$q := \frac{2p}{2p-1}.$$

From (1.2) it follows that

$$\hat{\delta}(H) := \inf_m \delta_m(H) = \inf_{j \neq k; j,k=1,2,\ldots} |\lambda_j(H) - \lambda_k(H)| > 0. \qquad (1.3)$$

Denote also

$$u_p(H) := 2N_{2p}(H_I)\zeta_q(H) \sum_{m=0}^{p-1} \sum_{k=0}^{\infty} \frac{\tau_p^{kp+m}(H)}{\hat{\delta}^{kp+m}(H)\sqrt{k!}},$$

where $\tau_p(H)$ is the number defined as in Section 9.5 and satisfying the inequality $\tau_p(H) \leq 2(1 + 2p)N_{2p}(H_I)$.

Theorem 12.1. *Let conditions (1.1) and (1.2) be fulfilled. Then there are a boundedly invertible operator $T \in \mathcal{B}(\mathcal{H})$ and a normal operator D acting in $\mathcal{H}$, such that (0.1) holds. Moreover,*

$$\kappa_T \leq e^{2u_p(H)}. \qquad (1.4)$$

The proof of this theorem is divided into a series of lemmas which are presented in the next three sections. The theorem is sharp: if H is selfadjoint, then $u_p(H) = 0$ and we obtain $\kappa_T = 1$.

To illustrate Theorem 12.1, consider a boundedly invertible operator $H = S+K$, where $K \in SN_{2p}$ and S is a positive definite selfadjoint operator with a discrete spectrum, whose eigenvalues are different, increasing and

$$\lambda_{j+1}(S) - \lambda_j(S) \geq b_0 j^\alpha$$

$$(b_0 = const > 0; \alpha > 1/q = (2p-1)/(2p); j = 1, 2, \ldots). \qquad (1.5)$$

Hence it follows that $S^{-1} \in SN_r$ for some $r > 1$. Since H is boundedly invertible, and $H = S(I + S^{-1}H)$ we conclude that $I + S^{-1}H$ is boundedly invertible. Thus $N_r(H^{-1}) \leq N_r(S^{-1})\|(I + S^{-1}H)^{-1}\|$. Since S is selfadjoint, we have

$$\sup_k \inf_j |\lambda_k(H) - \lambda_j(S)| \leq \|K\|,$$

cf. [60] (see also Section 9.2). Thus, if

$$2\|K\| < \inf_j(\lambda_{j+1}(S) - \lambda_j(S)), \tag{1.6}$$

then $\hat{\delta}(H) \geq \inf_j(\lambda_{j+1}(S) - \lambda_j(S) - 2\|K\|)$ and (1.2) holds with $\zeta_q(H) \leq \zeta_q(S, K)$, where

$$\zeta_q(S, K) := \left[\sum_{j=1}^{\infty} \frac{1}{(\lambda_{j+1}(S) - \lambda_j(S) - 2\|K\|)^q}\right]^{1/q} < \infty.$$

Example 12.1. Consider in $L^2(0,1)$ the spectral problem

$$u^{(4)}(x) + (Ku)(x) = \lambda u(x) \ (\lambda \in \mathbf{C}, 0 < x < 1);$$

$$u(0) = u(1) = u''(0) = u''(1) = 0,$$

where $K \in SN_{2p}$ for an integer $p \geq 1$. So H is defined by

$$H = \frac{d^4}{dx^4} + K$$

with

$$Dom(H) = \{v \in L^2(0,1) : v^{(4)} \in L^2(0,1), \ v(0) = v(1)$$

$$= v''(0) = v''(1) = 0\}.$$

Take $S = d^4/dx^4$ with $Dom(S) = Dom(H)$. Then $\lambda_j(S) = \pi^4 j^4$ and $\lambda_{j+1}(S) - \lambda_j(S) \geq 4\pi^4 j^3$ $(j = 1, 2, ...)$. If $\|K\| < 2\pi^4$, then H is boundedly invertible,

$$\hat{\delta}(H) \geq 4\pi^4 - 2\|K\|$$

and

$$\zeta_q^q(H) \leq \sum_{j=1}^{\infty} \frac{1}{(4\pi^4 j^3 - 2\|K\|)^q} < \infty.$$

Now one can directly apply Theorem 12.1.

12.2 Operators with finite invariant chains

Let B_0 be a bounded linear operator in $\mathcal{H}$ having a finite chain of invariant projections P_k $(k = 1, ..., n; \; n < \infty)$:

$$0 \subset P_1 \mathcal{H} \subset P_2 \mathcal{H} \subset ... \subset P_n \mathcal{H} = \mathcal{H} \tag{2.1}$$

and

$$P_k B_0 P_k = B_0 P_k \quad (k = 1, ..., n). \tag{2.2}$$

Put $\Delta P_k = P_k - P_{k-1}$ $(P_0 = 0)$ and denote by A_k the restriction of the operator $\Delta P_k B_0 \Delta P_k$ onto $\Delta P_k \mathcal{H}$ $(k = 1, ..., n)$.

Lemma 12.1. *Let conditions (2.1) and (2.2) hold. Then*

$$\sigma(B_0) = \cup_{k=1}^n \sigma(A_k).$$

Proof. Put

$$\hat{D} = \sum_{k=1}^n A_k \text{ and } W = B_0 - \hat{D}.$$

Due to (2.2) we have $W P_k = P_{k-1} W P_k$. Hence,

$$W^n = W^n P_n = W^{n-1} P_{n-1} W P_n = W^{n-2} P_{n-2} W P_{n-1} W P_n =$$

$$W^{n-2} P_{n-2} W^2 = W^{n-3} P_{n-3} W^3 = ... = P_0 W^n = 0.$$

So W is nilpotent. Similarly, taking into account that

$$(\hat{D} - \lambda I)^{-1} W P_k = (\hat{D} - \lambda I)^{-1} P_{k-1} W P_k = P_{k-1} (\hat{D} - \lambda I)^{-1} W P_k$$

we prove that $((\hat{D} - \lambda I)^{-1} W)^n = 0$ $(\lambda \notin \sigma(D))$. Thus

$$(B_0 - \lambda I)^{-1} = (\hat{D} + W - \lambda I)^{-1} = (I + (\hat{D} - \lambda I)^{-1} W)^{-1} (\hat{D} - \lambda I)^{-1} =$$

$$\sum_{k=0}^{n-1} (-1)^k ((\hat{D} - \lambda I)^{-1} W)^k (\hat{D} - \lambda I)^{-1}.$$

Hence it easily follows that $\sigma(\hat{D}) = \sigma(B_0)$. This proves the lemma, since A_k are mutually orthogonal. Q. E. D.

Under conditions (2.1), (2.2) denote

$$Q_k = I - P_k, B_k = Q_k B_0 Q_k \text{ and } C_k = \Delta P_k B_0 Q_k.$$

Since B_j is a block triangular operator matrix, according to the previous lemma we have

$$\sigma(B_j) = \cup_{k=j+1}^{n}\sigma(A_k) \quad (j = 0, ..., n).$$

It is assumed that the spectra $\sigma(A_k)$ of A_k satisfy the condition

$$\sigma(A_k) \cap \sigma(A_j) = \emptyset \quad (j \neq k; \; j, k = 1, ..., n). \tag{2.3}$$

Under this condition, according to the Rosenblum theorem the equation

$$A_j X_j - X_j B_j = -C_j \quad (j = 1, ..., n-1) \tag{2.4}$$

has a unique solution. We need the following result.

Lemma 12.2. *Let condition (2.3) hold and X_j be a solution to (2.4). Then*

$$(I - X_{n-1})(I - X_{n-2}) \cdots (I - X_1) B_0 (I + X_1)(I + X_2) \cdots (I + X_{n-1}) =$$

$$A_1 + A_2 + ... + A_n. \tag{2.5}$$

The proof of this lemma is similar to the proof of Lemma 6.3.

Take

$$\hat{T}_n = (I + X_1)(I + X_2) \cdots (I + X_{n-1}). \tag{2.6}$$

It is simple to see that $X_j^2 = 0$ and the inverse to $I + X_j$ is the operator $I - X_j$. Thus,

$$\hat{T}_n^{-1} = (I - X_{n-1})(I - X_{n-2}) \cdots (I - X_1) \tag{2.7}$$

and (2.5) can be written as

$$\hat{T}_n^{-1} B_0 \hat{T}_n = diag \, (A_k)_{k=1}^{n}. \tag{2.8}$$

By the inequalities between the arithmetic and geometric means we get

$$\|\hat{T}_n\| \leq \prod_{k=1}^{n-1}(1 + \|X_k\|) \leq \left(1 + \frac{1}{n-1}\sum_{k=1}^{n-1}\|X_k\|\right)^{n-1} \tag{2.9}$$

and

$$\|\hat{T}_n^{-1}\| \leq \left(1 + \frac{1}{n-1}\sum_{k=1}^{n-1}\|X_k\|\right)^{n-1}. \tag{2.10}$$

12.3 The finite dimensional case

In Chapter 6 we have obtained a bound for the condition number of a finite matrix in terms of the Hilbert-Schmidt norm. In the present section the similar result is derived in terms of the Schatten-von Nemann norm. In addition, we formulate our results in this section in the form which enables us to generalize them to the infinite dimensional case.

We are going to apply Lemma 12.2 to an $n \times n$-matrix A whose eigenvalues are different and are enumerated in the increasing way of their absolute values. So

$$\hat{\delta}(A) := \min_{j,k=1,\ldots,n;\ k\neq j} |\lambda_j(A) - \lambda_k(A)| > 0. \tag{3.1}$$

Consequently, there is an invertible matrix $T_n \in \mathbf{C}^{n\times n}$ and a normal matrix $D_n \in \mathbf{C}^{n\times n}$, such that

$$T_n^{-1}AT_n = D_n. \tag{3.2}$$

Let $\{e_k\}$ be the Schur basis in which A has the form

$$A = \begin{pmatrix} a_{11} & a_{12} & a_{13} & \cdots & a_{1n} \\ 0 & a_{22} & a_{23} & \cdots & a_{2n} \\ . & . & . & \cdots & . \\ 0 & 0 & 0 & \cdots & a_{nn} \end{pmatrix}$$

with $a_{jj} = \lambda_j(A)$. Take $P_j = \sum_{k=1}^{j}(\cdot,e_k)e_k$, $B_0 = A$, $\Delta P_k = (.,e_k)e_k$,

$$Q_j = \sum_{k=j+1}^{n} (.,e_k)e_k,\quad A_k = \Delta P_k A \Delta P_k = \lambda_k(A)\Delta P_k,$$

$$B_j = Q_j A Q_j = \begin{pmatrix} a_{j+1,j+1} & a_{j+1,j+2} & \cdots & a_{j+1,n} \\ 0 & a_{j+2,j+2} & \cdots & a_{j+2,n} \\ . & . & . & \cdots \\ 0 & 0 & . & a_{nn} \end{pmatrix},$$

$$C_j = \Delta P_j A Q_j = \begin{pmatrix} a_{j,j+1} & a_{j,j+2} & \cdots & a_{j,n} \end{pmatrix} \tag{3.3}$$

and

$$D_n = diag(\lambda_k(A)).$$

In addition,

$$A = \begin{pmatrix} \lambda_1(A) & C_1 \\ 0 & B_1 \end{pmatrix}, B_1 = \begin{pmatrix} \lambda_2(A) & C_2 \\ 0 & B_2 \end{pmatrix}, \ldots, B_j = \begin{pmatrix} \lambda_{j+1}(A) & C_{j+1} \\ 0 & B_{j+1} \end{pmatrix}$$

$(j < n)$. So B_j is an upper-triangular $(n - j) \times (n - j)$-matrix. Equation (2.4) takes the form

$$\lambda_j(A)X_j - X_j B_j = -C_j. \tag{3.4}$$

Since $X_j = X_j Q_j$, we can write $X_j(\lambda_j(A)Q_j - B_j) = C_j$. Therefore

$$X_j = C_j \, (\lambda_j(A)Q_j - B_j)^{-1}. \tag{3.5}$$

The inverse operator is understood in the sense of subspace $Q_j \mathbf{C}^n$. Hence,

$$\|X_j\| \leq \|C_j\| \|(\lambda_j(A)Q_j - B_j)^{-1}\|.$$

Besides, due to Theorem 9.5,

$$\|(\lambda_j(A)Q_j - B_j)^{-1}\| \leq \sum_{m=0}^{p-1} \sum_{k=0}^{\infty} \frac{\tau_p^{kp+m}(B_j)}{\delta_j^{kp+m+1}(A)\sqrt{k!}},$$

where

$$\delta_j(A) = \inf_{m=1,2,\ldots,n; \, m\neq j} |\lambda_j(A) - \lambda_m(A)| \quad (j \leq n).$$

But $N_{2p}(\Im B_j) = N_{2p}(Q_j A_I Q_j) \leq N_{2p}(A_I) \; (j \geq 1)$. Recall that $A_I = \Im A$. So $\tau_p(B_j) \leq \tau_p(A)$ and

$$\|(\lambda_j(A)Q_j - B_j)^{-1}\| \leq \frac{\eta(A)}{\delta_j(A)},$$

where

$$\eta(A) = \sum_{m=0}^{p-1} \sum_{k=0}^{\infty} \frac{\tau^{kp+m}(A)}{\hat{\delta}^{kp+m}(A)\sqrt{k!}} \quad (\hat{\delta}(A) = \inf_j \delta_j(A)).$$

Consequently,

$$\|X_j\| \leq \eta(A)\frac{\|C_j\|}{\delta_j(A)}$$

and

$$\sum_{j=1}^{n-1} \|X_j\| \leq \eta(A) \sum_{j=1}^{n-1} \frac{\|C_j\|}{\delta_j(A)}.$$

Hence, by the Hólder inequality,

$$\|X_j\| \leq \eta(A) \sum_{j=1}^{n-1} \frac{\|C_j\|}{\delta_j(A)} \leq \eta(A) \left(\sum_{j=1}^{n-1} \|C_j\|^{2p} \right)^{1/2p} \zeta_{q,n}(A) \quad (\frac{1}{2p} + \frac{1}{q} = 1), \tag{3.6}$$

where

$$\zeta_{q,n}(A) := \left(\sum_{k=1}^{n-1} \frac{1}{\delta_k^q(A)} \right)^{1/q}.$$

In addition,

$$\|C_j\|^2 = \|C_j^*\|^2 = \sum_{k=j+1}^{n} |a_{jk}|^2, j < n; \; C_n = 0,$$

and

$$4\|A_I e_j\|^2 = \|(A - A^*)e_j\|^2 = |a_{jj} - \overline{a}_{jj}|^2 + \sum_{k=j+1}^{n} |a_{jk}|^2 + \sum_{k=2}^{j-1} |a_{jk}|^2$$

$$\geq \|C_j\|^2 \; (j < n).$$

Thus, $\|C_j\| \leq 2\|A_I e_j\|$ and therefore

$$\sum_{j=1}^{n-1} \|C_j\|^{2p} \leq 2^p \sum_{j=1}^{n-1} \|A_I e_j\|^{2p}.$$

But according to the well known Lemmas II.4.1 and II.2.4 from [54], we deduce that

$$\sum_{j=1}^{n-1} \|A_I e_j\|^{2p} \leq N_{2p}^{2p}(A_I).$$

Therefore (3.6) implies

$$\sum_{j=1}^{n-1} \|X_j\| \leq \eta(A)\zeta_{q,n}(A)2N_{2p}(A_I). \tag{3.7}$$

Take $T_n = \hat{T}_n$ as in (2.6) with X_k defined by (3.5). Besides (2.9) and (2.10) imply

$$\|T_n\| \leq \prod_{k=1}^{n-1}(1 + \|X_k\|) \leq \left(1 + \frac{1}{n-1}\sum_{j=1}^{n-1}\|X_j\| \right)^{n-1} \leq \psi_{n,p}(A), \tag{3.8}$$

where

$$\psi_{n,p}(A) := \left(1 + \frac{2\eta(A)N_{2p}(A_I)\zeta_q(A)}{n-1} \right)^{n-1}.$$

Similarly,

$$\|T_n^{-1}\| \leq \psi_{n,p}(A). \tag{3.9}$$

We thus have proved the following result.

Lemma 12.3. *Let condition (3.1) be fulfilled. Then there is an invertible operator T_n, such that (3.2) holds with $\kappa_{T_n} = \|T_n^{-1}\|\|T_n\| \leq \psi_{n,p}^2(A)$.*

12.4 Proof of Theorem 12.1

We need the following result.

Lemma 12.4. *Under the hypothesis of Theorem 12.1, operator H^{-1} has a complete system of root vectors.*

Proof. We can write $H = H_R + iH_I$ with the notation $H_R = (H + H^*)/2$. For any real c with $-c \notin \sigma(H) \cup \sigma(H_R)$ we have

$$(H + cI)^{-1} = (I + i(H_R + cI)^{-1}H_I)^{-1}(H_R + cI)^{-1}.$$

But $(I + i(H_R + cI)^{-1}H_I)^{-1} - I = K_0$, where

$$K_0 = -i(H_R + cI)^{-1}H_I(I + i(H_R + cI)^{-1}H_I)^{-1}$$

is compact. So

$$(H + cI)^{-1} = (H_R + cI)^{-1}(I + K_0). \tag{4.1}$$

Due to (1.1) $(H + cI)^{-1} = H^{-1}(I + cH^{-1})^{-1} \in SN_r$. Hence,

$$(H_R + cI)^{-1} = (I + H - H_I i + cI)^{-1}$$

$$= (I - i(H + cI)^{-1}H_I)(H + cI)^{-1} \in SN_r$$

and therefore by (4.1) and the Keldysh Theorem 11.3 operator $(H + cI)^{-1}$ has a complete system of roots vectors. Since $(H + cI)^{-1}$ and H^{-1} commute, H^{-1} has a complete system of roots vectors, as claimed. Q. E. D.

From the previous lemma it follows that there is an orthonormal (Schur) basis $\{\hat{e}_k\}_{k=1}^{\infty}$, in which H^{-1} is represented by a triangular matrix (see [54, Lemma I.4.1]). Denote $\hat{P}_k = \sum_{j=1}^{k}(., \hat{e}_j)\hat{e}_j$. Then

$$H^{-1}\hat{P}_k = \hat{P}_k H^{-1}\hat{P}_k \quad (k = 1, 2, ...).$$

Besides,

$$\Delta\hat{P}_k H^{-1}\Delta\hat{P}_k = \lambda_k^{-1}(H)\Delta\hat{P}_k \quad (\Delta\hat{P}_k = \hat{P}_k - \hat{P}_{k-1}, \ k = 1, 2, ...; \hat{P}_0 = 0). \tag{4.2}$$

Put

$$D = \sum_{k=1}^{\infty}\lambda_k\Delta\hat{P}_k \quad (\Delta\hat{P}_k(H) = \hat{P}_k - \hat{P}_{k-1}, \ k = 1, 2, ...) \text{ and } V = H - D.$$

We have

$$H\hat{P}_k f = \hat{P}_k H\hat{P}_k f \ (k = 1, 2, ...; \ f \in Dom(H)). \tag{4.3}$$

Indeed, $H^{-1}\hat{P}_k$ is an invertible $k \times k$ matrix, and therefore, $H^{-1}\hat{P}_k\mathcal{H}$ is dense in $\hat{P}_k\mathcal{H}$. Since $\Delta\hat{P}_j\hat{P}_k = 0$ for $j > k$, we have $0 = \Delta\hat{P}_jHH^{-1}\hat{P}_k = \Delta\hat{P}_jH\hat{P}_kH^{-1}\hat{P}_k$. Hence $\Delta\hat{P}_jHf = 0$ for any $f \in \hat{P}_kH$. This implies (4.3).

Furthermore, put $H_n = H\hat{P}_n$. Due to (4.3) we have

$$\|H_nf - Hf\| \to 0 \ (f \in Dom(H)) \text{ as } n \to \infty. \tag{4.4}$$

For an integer $j \geq 2$ put

$$B_j(H) = (I - \hat{P}_j)H(I - \hat{P}_j), C_j(H) = \Delta\hat{P}_jH(I - \hat{P}_j),$$

$$X_j(H) = C_j(H) \left(\lambda_j(H)(I - \hat{P}_j) - B_j(H)\right)^{-1}$$

and

$$T_n(H) = \prod_{1 \leq k \leq n-1}^{\rightarrow} (I + X_k(H)).$$

Due to (3.8)

$$\|T_n(H)\| \leq \prod_{1 \leq k \leq n-1}^{\rightarrow} (1 + \|X_k(H)\|) \leq \left(1 + \frac{2\eta(H)N_{2p}(H_I)\zeta_q(H)}{n-1}\right)^{n-1}$$

$$= \left(1 + \frac{u_p(H)}{n-1}\right)^{n-1} \leq exp\left(u_p(H)\right). \tag{4.5}$$

Hence it follows that the product

$$T(H) := \prod_{1 \leq k \leq \infty}^{\rightarrow} (1 + X_k(H))$$

converges in the operator norm and $T_n(H)\hat{P}_n \to T(H)$. But due to Lemma 12.3 $T_n(H)D\hat{P}_n = HP_nT_n(H)$. Letting $n \to \infty$ we get $T(H)Df = HT(H)f$ ($f \in Dom(H)$). In addition, by (4.5) $\|T(H)\| \leq e^{u_p(H)}$. Similarly, $\|T^{-1}(H)\| \leq e^{u_p(H)}$. This finishes the proof. Q. E. D.

12.5 Condition numbers of boundedly perturbed normal operators

In this section we consider a closed linear operator A in a separable Hilbert space with the following property: there is a normal operator D with a discrete spectrum, such that $Dom(D) = Dom(A)$ and

$$\nu := \|A - D\| < \infty. \tag{5.1}$$

Under certain assumptions it is shown that A is similar to a normal operator and a sharp bound for the condition number is derived. The approach suggested in this section is considerably different from the one considered in Sections 12.1-12.4.

It is assumed that all the eigenvalues of D are different:

$$d_m := \inf_{j=1,2,\ldots;\, j \neq m} |\lambda_j(D) - \lambda_m(D)|/2 > 0 \text{ and } \hat{d} := \inf_m d_m > 0. \quad (5.2)$$

Now we are in a position to formulate our next result.

Theorem 12.2. *Let the conditions (5.1), (5.2),*

$$2\nu < \hat{d} \quad (5.3)$$

and

$$\sum_{k=1}^{\infty} \frac{1}{(d_k - 2\nu)^2} < \infty \quad (5.4)$$

hold. Then there are a boundedly invertible operator $T \in \mathcal{B}(\mathcal{H})$, and a normal operator S, acting in $\mathcal{H}$, such that

$$TAx = STx \quad (x \in Dom(A)). \quad (5.5)$$

Moreover,

$$\kappa_T \leq \gamma_0(A), \text{ where } \gamma_0(A) := \frac{\hat{d}}{\hat{d} - \nu} \left[1 + 2\nu \left(\sum_{k=1}^{\infty} \frac{1}{(d_k - 2\nu)^2} \right)^{1/2} \right]^2.$$

The proof of this theorem is presented in the next section. The theorem is sharp: if $A = D$ is normal, then $\nu = 0$ and we obtain $\gamma_0(A) = 1$.

Example 12.2. Consider in $L^2(0,1)$ the problem

$$-u''(x) + p(x)u(x) = \lambda u(x) \ (0 < x < 1); \ u(0) = u(1) = 0, \quad (5.6)$$

where $p(x)$ is a bounded measurable complex valued function. So A is defined by $(Au)(x) = -u''(x) + p(x)u(x)$ with

$$Dom(A) = \{v \in L^2(0,1) : v'' \in L^2(0,1), \ v(0) = v(1) = 0\}.$$

Let $\hat{p}(x) = p(x) - p_0 \ (0 \leq x \leq 1)$ for some constant p_0. Take $D = -d^2/dx^2 + p_0$ with $Dom(D) = Dom(A)$. Then $\lambda_j(D) = \pi^2 j^2 + p_0 \ (j = 1, 2, \ldots)$ and $d_j = (\lambda_{j+1}(D) - \lambda_j(D))/2 = \pi^2(j + 1/2)$, $\hat{d} = 3\pi^2/2$. In addition,

$$\nu = \|A - D\| = \sup_x |\hat{p}(x)|.$$

If $4\nu < 3\pi^2$, then

$$\sum_{j=1}^{\infty} \frac{1}{(\pi^2(j+1/2) - 2\nu)^2} < \infty.$$

So condition (5.4) holds. Now one can directly apply Theorem 12.2.

Let us compare Theorem 12.2 with the well known Theorem 5.4.15a from [60] which asserts that the perturbed operator remains a scalar type spectral operator under considerations but, in contrast to the above stated Theorem 12.2, Theorem 5.4.15a from [60] does not give a bound for the condition number in the case of the simple spectrum.

12.6   Proof of Theorem 12.2

Let $\Omega(a, r) = \{z \in \mathbf{C} : |z - a| \leq r\}$ $(a \in \mathbf{C}, r > 0)$ and $\{e_k\}$ be the set of all the (mutually orthogonal) normed eigenvectors of D. So

$$D = \sum_{k=1}^{\infty} \lambda_k(D)P_k, \text{ where } P_k = (., e_k)e_k.$$

Denote by Q_m the Riesz projection of A corresponding to the eigenvalues of A lying in $\Omega(\lambda_m(D), d_m)$ and suppose that (5.1) and (5.2) hold. Since D is normal, we have $\|R_\lambda(D)\| = 1/\rho(D, \lambda)$. Thus $\lambda \notin \sigma(A)$, provided $\nu < \rho(D, \lambda)$. Hence it follows that

$$\sup_{s \in \sigma(A)} \inf_{t \in \sigma(D)} |s - t| \leq \nu.$$

In the selfadjoint case this inequality is given in [60, p. 291]. Consequently,

$$\sigma(A) \subset \cup_{m=1}^{\infty} \Omega(\lambda_m(D), \nu).$$

Making use of Lemma 1.12 with $A = D$ and $\tilde{A} = A$, taking into account that $\|R_\lambda(D)\| = 1/\rho(D, \lambda)$, under condition (5.3), we have

$$\|P_m - Q_m\| \leq \delta_m, \text{ where } \delta_m := \frac{\nu}{d_m - \nu} < 1. \qquad (6.1)$$

Let $\{g_k\}$ be the set of all the eigenvectors of A and $\{h_k\}$ the corresponding biorthogonal sequence: $(g_k, h_j) = 0, k \neq j$, $(g_k, h_k) = 1$ $(j, k = 1, 2, ...)$, $(h_k$ are the eigenvectors of A^*). Then $Q_k = (., h_k)g_k$ and

$$A = \sum_{k=1}^{\infty} \lambda_k(A)Q_k.$$

Put

$$T = \sum_{k=1}^{\infty} (., h_k) e_k. \tag{6.2}$$

Simple calculations show that the inverse operator is defined by

$$T^{-1} = \sum_{k=1}^{\infty} (., e_k) g_k. \tag{6.3}$$

Below we check that T and T^{-1} are bounded.

Lemma 12.5. *Let conditions (5.1)-(5.3) hold and T be defined by (6.2). Then (5.5) is valid with*

$$S = \sum_{k=1}^{\infty} \lambda_k(A) P_k. \tag{6.4}$$

Proof. Indeed,

$$AT^{-1}f = \sum_{k=1}^{\infty} \sum_{j=1}^{\infty} \lambda_k(A)(f, e_j)(g_j, h_k) g_k = \sum_{k=1}^{\infty} \lambda_k(A)(f, e_k) g_k$$

$$(T^{-1}f \in Dom(A))$$

and

$$TAT^{-1}f = \sum_{k=1}^{\infty} \lambda_k(A) \sum_{j=1}^{\infty} (g_k, h_j) e_j (f, e_k) = \sum_{k=1}^{\infty} \lambda_k(A)(f, e_k) e_k = Sf,$$

as claimed. Q. E. D.

Introduce the operator

$$J = \sum_{k=1}^{\infty} \|h_k\| (., e_k) e_k.$$

Then for any $f \in \mathcal{H}$ we have

$$Tf - Jf = \sum_{k=1}^{\infty} \|h_k\| (f, \hat{h}_k - e_k) e_k, \text{ where } \hat{h}_k = h_k / \|h_k\|.$$

Hence,

$$\|Tf - Jf\|^2 = \sum_{k=1}^{\infty} \|h_k\|^2 |(f, \hat{h}_k - e_k)|^2 \leq \|f\|^2 \sum_{k=1}^{\infty} \|h_k\|^2 \|\hat{h}_k - e_k\|^2. \tag{6.5}$$

It is clear the h_k are the eigenvectors of A^*. Besides $\|A^* - D^*\| = \|A - D\| = \nu$.

Applying Lemma 1.13, with $\tilde{A} = A^*$ and $A = D^*$, according to (6.1) we can write

$$\|e_m - \hat{h}_m\| \leq \frac{2\delta_m}{1 - \delta_m} = \frac{2\nu}{d_m - 2\nu}.$$

Now (6.5) implies

$$\|T - J\|^2 \leq (2\nu)^2 \sum_{k=1}^{\infty} \frac{\|h_k\|^2}{(d_k - 2\nu)^2}. \tag{6.6}$$

We can take h_k and g_k in such a way that $\|h_k\| = \|g_k\|$. Clearly, $Q_k h_k = (h_k, h_k) g_k$. So

$$(Q_k h_k, g_k) = (h_k, h_k)(g_k, g_k) = \|h_k\|^4 = \|g_k\|^4.$$

Hence,

$$\|h_k\|^4 \leq \|Q_k\| \|h_k\| \|g_k\| = \|Q_k\| \|h_k\|^2.$$

Thus

$$\|h_k\|^2 \leq \|Q_k\| \quad \text{and} \quad \|g_k\|^2 \leq \|Q_k\|. \tag{6.7}$$

Now (6.6) implies

$$\|T - J\|^2 \leq (2\nu)^2 \sum_{k=1}^{\infty} \frac{\|Q_k\|}{(d_k - 2\nu)^2}.$$

Moreover, by (6.1),

$$\|Q_k\| \leq \|P_k\| + \frac{\nu}{d_k - \nu} = 1 + \frac{\nu}{d_k - \nu} \leq c_0^2 \quad (k = 1, 2, ...), \tag{6.8}$$

where

$$c_0 = \left(1 + \frac{\nu}{\hat{d} - \nu}\right)^{1/2} = \frac{\sqrt{\hat{d}}}{\sqrt{\hat{d} - \nu}}.$$

Consequently,

$$\|T - J\|^2 \leq (2\nu c_0)^2 \sum_{k=1}^{\infty} \frac{1}{(d_k - 2\nu)^2}.$$

Hence,

$$\|T\| \leq \|J\| + \|T - J\| \leq \|J\| + 2c_0\nu \left[\sum_{k=1}^{\infty} \frac{1}{(d_k - 2\nu)^2}\right]^{1/2}.$$

But due to (6.7) and (6.8),

$$\|Jf\|^2 = \sum_{k=1}^{\infty} \|h_k\|^2 |(f, e_k)|^2 \le c_0^2 \sum_{k=1}^{\infty} |(f, e_k)|^2 = \|f\|^2 c_0^2 \quad (f \in \mathcal{H}).$$

Thus we obtain

$$\|T\| \le c_0 \left(1 + 2\nu \left[\sum_{k=1}^{\infty} \frac{1}{(d_k - 2\nu)^2}\right]^{1/2}\right).$$

The same arguments along with (6.3) and (6.7) give us the inequality

$$\|T^{-1}\| \le c_0 \left(1 + 2\nu \left[\sum_{k=1}^{\infty} \frac{1}{(d_k - 2\nu)^2}\right]^{1/2}\right).$$

This finishes the proof. Q. E. D.

12.7 Applications of condition numbers

Rewrite (0.1) as $Hx = T^{-1}DTx$. Let ΔP_k be the eigenprojections of the normal operator D and $E_k = T^{-1}\Delta P_k T$. Then

$$Hx = T^{-1} \sum_{k=1}^{\infty} \lambda_k(H)\Delta P_k Tx \quad (x \in Dom(H)).$$

Let $f(z)$ be a scalar function defined and bounded on the spectrum of H. Put

$$f(H) = \sum_{k=1}^{\infty} f(\lambda_k(H)) E_k.$$

Then

$$f(H) = T^{-1} \sum_{k=1}^{\infty} f(\lambda_k(H))\Delta P_k T = T^{-1} f(D) T.$$

So we arrive at the following result.

Corollary 12.1. *If condition (0.1) holds and* $\sup_k |f(\lambda_k(H))| < \infty$, *then* $\|f(H)\| \le \kappa_T \sup_k |f(\lambda_k(H))|$.

In particular, we have

$$\|e^{-Ht}\| \le \kappa_T(H)e^{-\beta(H)t} \quad (t \ge 0),$$

where $\beta(H) = \inf_k \Re \lambda_k(H)$. In addition,

$$\|R_\lambda(H)\| \le \frac{\kappa_T}{\rho(H,\lambda)} \quad (\lambda \notin \sigma(H)). \tag{7.1}$$

Recall that

$$sv_A(\tilde{A}) := \sup_{t \in \sigma(\tilde{A})} \inf_{s \in \sigma(A)} |t - s|$$

is the variation of the spectrum of $\tilde{A}$ with respect to the spectrum of A.

Now let $\tilde{H}$ be a linear operator in $\mathcal{H}$ with $Dom(\tilde{H}) = Dom(\tilde{H})$ and

$$\xi := \|H - \tilde{H}\| < \infty. \tag{7.2}$$

From (7.1) it follows that $\lambda \notin \sigma(\tilde{H})$, provided $\xi\kappa_T < \rho(H,\lambda)$. So for any $\mu \in \sigma(\tilde{H})$ we have $\xi\kappa_T \ge \rho(H,\mu)$. This inequality implies our next result.

Corollary 12.2. *Let conditions (0.1) and (7.2) hold. Then* $sv_H(\tilde{H}) \le \xi\kappa_T$.

Now consider unbounded perturbations assuming that H is boundedly invertible. To this end put

$$H^\theta = \sum_{k=1}^\infty \lambda_k^\theta(H)E_k \quad (0 < \theta < 1).$$

We have

$$\|H^\theta R_\lambda(H)\| \le \frac{\kappa_T}{\phi_\theta(H,\lambda)} \quad (\lambda \notin \sigma(H)), \tag{7.3}$$

where

$$\phi_\theta(H,\lambda) := \inf_k |(\lambda - \lambda_k(H))\lambda_k^{-\theta}(H)|.$$

Let $\tilde{H}$ be a linear operator in $\mathcal{H}$ with $Dom(H) = Dom(\tilde{H})$ and

$$\xi_\theta := \|(H - \tilde{H})H^{-\theta}\| < \infty. \tag{7.4}$$

Take into account that

$$R_\lambda(H) - R_\lambda(\tilde{H}) = R_\lambda(H)(\tilde{H} - H)R_\lambda(\tilde{H})$$

$$= R_\lambda(\tilde{H})(\tilde{H} - H)H^{-\theta}H^\theta R_\lambda(H) \quad (\lambda \notin \sigma(H) \cup \sigma(\tilde{H})).$$

Therefore,

$$\|R_\lambda(\tilde{H})\| \le \|R_\lambda(H)\| + \|R_\lambda(\tilde{H})\|\|(\tilde{H} - H)H^{-\theta}\|\|H^\theta R_\lambda(H)\|.$$

Hence and from (7.3) it follows that $\lambda \notin \sigma(\tilde{H})$, provided $\lambda \notin \sigma(H)$, and the conditions (7.4) and $\xi_\theta \kappa_T < \phi_\theta(H, \lambda)$ hold. So for any $\mu \in \sigma(\tilde{H})$ we have

$$\xi_\theta \kappa_T \geq \phi_\theta(H, \mu). \tag{7.5}$$

The quantity

$$\theta - \mathrm{rsv}_H(\tilde{H}) := \sup_{t \in \sigma(\tilde{H})} \inf_{s \in \sigma(H)} |(t - s)s^{-\theta}| = \sup_{t \in \sigma(\tilde{H})} \phi_\theta(H, t)$$

is said to be the θ-relative spectral variation of operator $\tilde{H}$ with respect to H. Now (7.5) implies.

Corollary 12.3. *Let conditions (0.1) and (7.4) hold. Then* $\theta - \mathrm{rsv}_H(\tilde{H}) \leq \xi_\theta \kappa_T$.

12.8   Comments to Chapter 12

This chapter is based on the papers [40] and [42].

Chapter 13

Commutators and Perturbations of Operator Functions

Let $\mathcal{X}$ be a Banach space. For $A, B, \tilde{A} \in \mathcal{B}(\mathcal{X})$, $[A, B] := AB - BA$ is the commutator, $[A, B, \tilde{A}] := AB - B\tilde{A}$ is the generalized commutator; $[f(A), B] := f(A)B - Bf(A)$ and $[f(A), B, f(\tilde{A})] := f(A)B - Bf(\tilde{A})$ will be called the function commutator and the generalized function commutator, respectively.

In this chapter we derive norm estimates for $[f(A), B, f(\tilde{A})]$. In the case $B = I$ these estimates give us the bounds for $\|f(A) - f(\tilde{A})\|$.

13.1 Representations of commutators

We begin with the following lemma

Lemma 13.1. *Let* $A, \tilde{A}, B \in \mathcal{B}(\mathcal{X})$. *Then for any* $z \notin \sigma(A) \cup \sigma(\tilde{A})$, *we have*

$$(zI - A)^{-1}B - B(zI - \tilde{A})^{-1} = (Iz - A)^{-1}K(Iz - \tilde{A})^{-1}, \qquad (1.1)$$

where

$$K := AB - B\tilde{A} = [A, B, \tilde{A}].$$

Proof. Multiplying the both sides of (1.1) by $zI - A$ from the left and by $zI - \tilde{A}$ from the right, we have

$$B(zI - \tilde{A}) - (zI - A)B = K.$$

This proves the lemma. Q. E. D.

Lemma 13.2. *Let* $A, \tilde{A}, B \in \mathcal{B}(\mathcal{X})$. *Let* $f(z)$ *be regular on a neighborhood of* $\sigma(A) \cup \sigma(\tilde{A})$. *Then*

$$f(A)B - Bf(\tilde{A}) = \frac{1}{2\pi i} \int_L f(z) R_z(A) K R_z(\tilde{A}) dz, \qquad (1.2)$$

where L a closed smooth contour surrounding $\sigma(A) \cup \sigma(\tilde{A})$.

Proof. Lemma 13.1 and (1.2) imply

$$f(A)B - Bf(\tilde{A}) = -\frac{1}{2\pi i} \int_L f(z)(R_z(A)B - BR_z(\tilde{A}))dz =$$

$$\frac{1}{2\pi i} \int_L f(z)R_z(A)KR_z(\tilde{A})dz,$$

as claimed. Q. E. D.

Put $\Omega(r) = \{|z| \leq r\}$ and $\partial\Omega(r) = \{|z| = r\}$. In the rest of this section it is assumed that $f(z)$ is regular on $\Omega(r)$ with

$$r > r_s(A, \tilde{A}) := \max\{r_s(A), r_s(\tilde{A})\}.$$

Take into account that

$$R_\lambda(A) = -\sum_{k=0}^{\infty} \frac{A^k}{\lambda^{k+1}} \quad (|\lambda| > r_s(A)).$$

Then by the previous lemma

$$f(A)B - Bf(\tilde{A}) = \frac{1}{2\pi i} \int_{\partial\Omega(r)} f(z)R_z(A)KR_z(\tilde{A})dz$$

$$= \sum_{j,k=0}^{\infty} \frac{1}{2\pi i} \int_{\partial\Omega(r)} \frac{f(z)dz}{z^{k+j+2}} A^j K\tilde{A}^k.$$

Or

$$f(A)B - Bf(\tilde{A}) = \sum_{j,k=0}^{\infty} f_{j+k+1} A^j K\tilde{A}^k, \tag{1.3}$$

where f_j are the Taylor coefficients of f at zero. If, in particular, $f(z) = z^m$ for an integer $m \geq 1$, then we arrive at

Corollary 13.1. *Let $A, \tilde{A}, B \in \mathcal{B}(\mathcal{X})$. Then*

$$A^m B - B\tilde{A}^m = \sum_{j=0}^{m-1} A^j K\tilde{A}^{m-j-1} \quad (m = 2, 3, ...). \tag{1.4}$$

Take $f(z) = e^{zt}, t \geq 0$. Then the following result is true.

Lemma 13.3. *Let $A, B \in \mathcal{B}(\mathcal{X})$ and $K = [A, B]$. Then*

$$[e^{At}, B] = \int_0^t e^{As} K e^{A(t-s)} ds \quad (t \geq 0).$$

Proof. Put

$$J(t) = \int_0^t e^{As} K e^{A(t-s)} ds.$$

Then

$$\frac{d}{dt}(J(t)e^{-tA}) = e^{At} K e^{-At}.$$

On the other hand

$$\frac{d}{dt}([e^{At}, B]e^{-tA}) = \frac{d}{dt}(e^{At} B e^{-At} - B) = e^{At} K e^{-At}.$$

So $[e^{At}, B] = J(t)$ $(t > 0)$. In addition, for $t = 0$ we have $[I, B] = J(0) = 0$. This completes the proof. Q. E. D.

13.2 The finite dimensional case

Let $A, \tilde{A} \in \mathbf{C}^{n \times n}$ $(n < \infty)$. Recall that $g(A)$ is defined in Section 3.1. Denote by $co(A, \tilde{A})$ the closed convex hull of $\sigma(A) \cup \sigma(\tilde{A})$.

Now we are in a position to formulate the main result of this section.

Theorem 13.1. *Let* $A, \tilde{A}, B \in \mathbf{C}^{n \times n}$ *and* $f(\lambda)$ *be holomorphic on a neighborhood of* $co(A, \tilde{A})$. *Then with the notations*

$$\eta_{j,k} := \sup_{z \in co\,(A, \tilde{A})} \frac{|f^{(k+j+1)}(z)|}{\sqrt{k! j!}(k+j+1)!} \quad (j, k = 0, 1, 2, ...),$$

we have the inequality

$$N_2(f(A)B - Bf(\tilde{A})) \le N_2(K) \sum_{j,k=0}^{n-1} \eta_{j,k} g^j(A) g^k(\tilde{A}) \quad (K = AB - B\tilde{A}).$$

To prove this theorem we apply the triangular representation

$$A = D + V \quad (\sigma(A) = \sigma(D)), \tag{2.1}$$

where D is a normal and V is a nilpotent operators having the joint invariant subspaces. Similarly,

$$\tilde{A} = \tilde{D} + \tilde{V} \quad (\sigma(\tilde{A}) = \sigma(\tilde{D})), \tag{2.2}$$

where $\tilde{D}$ is a normal and $\tilde{V}$ is a nilpotent operators having joint invariant subspaces. The proof of Theorem 13.1 is based on the following lemma.

Lemma 13.4. *Under the hypothesis of Theorem 13.1, one has*

$$N_2(f(A)B - Bf(\tilde{A})) \le N_2(K) \sum_{j,k=0}^{n-1} \eta_{j,k} N_2^j(V) N_2^k(\tilde{V}). \tag{2.3}$$

Proof. By (2.1)

$$R_\lambda(A) = (D + V - I\lambda)^{-1} = (I + R_\lambda(D)V)^{-1} R_\lambda(D).$$

Note that $R_\lambda(D)V$ is a nilpotent matrix and therefore $(R_\lambda(D)V)^n = 0$. Consequently,

$$R_\lambda(A) = \sum_{k=0}^{n-1} (-1)^k (R_\lambda(D)V)^k R_\lambda(D).$$

Similarly,

$$R_\lambda(\tilde{A}) = \sum_{k=0}^{n-1} (-1)^k (R_\lambda(\tilde{D})\tilde{V})^k R_\lambda(\tilde{D}).$$

So by (1.2) we have

$$f(A)B - Bf(\tilde{A}) = \sum_{m,k=0}^{n-1} C_{mk} \tag{2.4}$$

where

$$C_{mk} = (-1)^{k+m} \frac{1}{2\pi i} \int_L f(\lambda)(R_\lambda(D)V)^m R_\lambda(D) K (R_\lambda(\tilde{D})\tilde{V})^k R_\lambda(\tilde{D})d\lambda.$$

Since D is a diagonal matrix in the orthonormal basis of the triangular representations of A (the Schur basis) $\{e_k\}$, and $\tilde{D}$ is a diagonal matrix in the Schur basis $\{\tilde{e}_k\}$ of $\tilde{A}$, we can write out

$$R_\lambda(D) = \sum_{j=1}^{n} \frac{Q_j}{\lambda_j - \lambda}, R_\lambda(\tilde{D}) = \sum_{j=1}^{n} \frac{\tilde{Q}_j}{\tilde{\lambda}_j - \lambda},$$

where $\lambda_j = \lambda_j(A), \tilde{\lambda}_j = \lambda_j(\tilde{A}), Q_k = (., e_k)e_k, \tilde{Q}_k = (., \tilde{e}_k)\tilde{e}_k$. Besides,

$$Q_j V Q_k = \tilde{Q}_j \tilde{V} \tilde{Q}_k = 0 \quad (j \geq k). \tag{2.5}$$

We can write,

$$C_{mk} = \sum_{i_1=1}^{n} Q_{i_1} V \sum_{i_2=1}^{n} Q_{i_2} V \dots V \sum_{i_{m+1}=1}^{n} Q_{i_{m+1}} K \sum_{j_1=1}^{n} \tilde{Q}_{j_1} \tilde{V} \sum_{j_2=1}^{n} \tilde{Q}_{j_2} \tilde{V} \dots$$

$$\tilde{V} \sum_{j_{k+1}=1}^{n} \tilde{Q}_{j_{k+1}} J_{i_1,i_2,\dots,i_{m+1},j_1 j_2 \dots j_{k+1}}. \tag{2.6}$$

Here

$$J_{i_1,i_2,\dots,i_{m+1},j_1 j_2 \dots j_{k+1}} =$$

$$\frac{(-1)^{k+m}}{2\pi i} \int_L \frac{f(\lambda)d\lambda}{(\lambda_{i_1} - \lambda)\dots(\lambda_{i_{m+1}} - \lambda)(\tilde{\lambda}_{j_1} - \lambda)\dots(\tilde{\lambda}_{j_{k+1}} - \lambda)}.$$

Below the symbol $|V|_e$ means the operator whose entries are absolute values of V in the basis $\{e_k\}$ and $|\tilde{V}|_{\tilde{e}}$ means the operator whose entries are absolute values of $\tilde{V}$ in the basis $\{\tilde{e}_k\}$. Furthermore, denote $K_{kj} = (K\tilde{e}_j, e_k)$ and $c_{kj}^{(ml)} = (C_{ml}\tilde{e}_j, e_k)$. Then

$$K = \sum_{j,k=1}^n K_{kj}(.,\tilde{e}_j)e_k \text{ and } C_{ml} = \sum_{j,k=1}^n c_{kj}^{(ml)}(.,\tilde{e}_j)e_k.$$

Put

$$|K|_{e\tilde{e}} = \sum_{j,k=1}^n |K_{kj}|(.,\tilde{e}_j)e_k \text{ and } |C_{ml}|_{e\tilde{e}} = \sum_{j,k=1}^n |c_{kj}^{(ml)}|(.,\tilde{e}_j)e_k.$$

By Lemma 3.8

$$|J_{i_1,i_2,\dots,i_{m+1},j_1j_2\cdots j_{k+1}}| \le \tilde{\eta}_{m,k} := \sup_{z \in co\,(A,\tilde{A})} \frac{|f^{(k+m+1)}(z)|}{(m+k+1)!}.$$

Now (2.6) and the equality

$$\sum_{k=1}^n Q_k = I$$

imply

$$|C_{mk}|_{e\tilde{e}}$$

$$\le \tilde{\eta}_{m,k} \sum_{i_1=1}^n Q_{i_1}|V|_e \sum_{i_2=1}^n Q_{i_2}|V|_e \cdots |V|_e \sum_{i_{m+1}=1}^n Q_{j_{m+1}}|K|_{e\tilde{e}} \sum_{j_1=1}^n \tilde{Q}_{j_2}|\tilde{V}|_{\tilde{e}} \cdots$$

$$\cdots |\tilde{V}|_{\tilde{e}} \sum_{j_{k+1}=1}^n \tilde{Q}_{j_{k+1}} = \tilde{\eta}_{m,k}|V|_e^m |K|_{e\tilde{e}}|\tilde{V}|_{\tilde{e}}^k. \tag{2.7}$$

The inequalities are understood in the entry-wise sense. Note that

$$N_2^2(|K|_{e\tilde{e}}) = \sum_{k=1}^n \||K|_{e\tilde{e}}\tilde{e}_k\|^2 = \sum_{k=1}^n \sum_{j=1}^n |K_{jk}|^2 = N_2^2(K).$$

Hence (2.7) yields the inequality

$$N_2(C_{mk}) \le \tilde{\eta}_{m,k}\||V|_e^m\|N_2(K)\||\tilde{V}|_{\tilde{e}}^k\|.$$

By Lemma 3.4,

$$\||V|_e^m\| \le \frac{N_2^m(|V|_e)}{\sqrt{m!}} = \frac{N_2^m(V)}{\sqrt{m!}}.$$

So

$$N_2(C_{mk}) \leq \tilde{\eta}_{m,k} N_2(K) \frac{N_2^m(V) N_2^k(\tilde{V})}{\sqrt{m!k!}}.$$

Now (2.4) implies the required result. Q. E. D.

Proof of Theorem 13.1: By Lemma 3.2 $N_2(V) = g(A)$. Now the required result is due to the preceding lemma. Q. E. D.

Taking in $B = I$, we get

Corollary 13.2. *Let A and $\tilde{A}$ be n-dimensional and $f(\lambda)$ be holomorphic on a neighborhood of $co(A, \tilde{A})$. Then*

$$N_2(f(A) - f(\tilde{A})) \leq N_2(A - \tilde{A}) \sum_{j,k=0}^{n-1} \eta_{j,k} g^j(A) g^k(\tilde{A}). \qquad (2.8)$$

13.3 Operators with Hilbert-Schmidt components

In the present section we consider the generalized function commutator with non-normal operators satisfying the conditions

$$A_I := (A - A^*)/2i \in SN_2, \tilde{A}_I := (\tilde{A} - \tilde{A}^*)/2i \in SN_2, \qquad (3.1)$$

and

$$K = AB - B\tilde{A} \in SN_2. \qquad (3.2)$$

Recall that $g_I(A)$ is defined in Section 9.2 and $g_I(A) \leq \sqrt{2} N_2(A_I)$. Again denote by $co(A, \tilde{A})$ the closed convex hull of $\sigma(A) \cup \sigma(\tilde{A})$ and

$$\eta_{j,k} := \sup_{z \in co(A, \tilde{A})} \frac{|f^{(k+j+1)}(z)|}{\sqrt{k!j!(k+j+1)!}} \quad (j, k = 0, 1, 2, ...).$$

Theorem 13.2. *Let conditions (3.1) and (3.2) hold. Let $f(\lambda)$ be holomorphic on a neighborhood of $co(A, \tilde{A})$. Then*

$$N_2(f(A)B - Bf(\tilde{A})) \leq N_2(K) \sum_{j,k=0}^{\infty} \eta_{j,k} g_I^j(A) g_I^k(\tilde{A}).$$

Proof. By Corollary 10.1 there are sequences $A_n, \tilde{A}_n$ $(n = 1, 2, ...)$ of finite dimensional operators strongly converging to A and $\tilde{A}$, respectively, such that $f(A_n) \overset{s}{\to} f(A)$ and $f(\tilde{A}_n) \overset{s}{\to} f(\tilde{A})$. Recall that the symbol $\overset{s}{\to}$ means the strong convergence. Note that

$$\|(f(A)B - Bf(\tilde{A}) - (f(A_n)B - Bf(\tilde{A}_n)))x\| \leq$$

$$\|(f(A) - f(A_n))Bx\| + \|B(f(\tilde{A}) - f(\tilde{A}_n))x\| \to 0 \quad (x \in \mathcal{H}).$$

Hence

$$f(A_n)B - Bf(\tilde{A}_n) \xrightarrow{s} [f(A), B, f(\tilde{A})].$$

But $[f(A), B, f(\tilde{A})] \in SN_2$ and therefore

$$N_2([f(A), B, f(\tilde{A})]) = \lim_{n \to \infty} N_2(f(A_n)B - Bf(\tilde{A}_n)). \tag{3.3}$$

Take rank $A_n = $ rank $\tilde{A}_n = n$. Then Theorem 13.1 yields

$$N_2(f(A_n)B - Bf(\tilde{A}_n)) \le N_2(A_n B - B\tilde{A}_n) \sum_{j,k=0}^{n-1} \eta_{j,k} g^j(A_n) g^k(\tilde{A}_n).$$

Moreover, Theorem 3.1 and Corollary 10.2 imply $g(A_n) = g_I(A_n) \to g_I(A)$. Now (3.3) proves the theorem. Q. E. D.

Taking in the previous theorem $B = I$ we get

$$N_2(f(A) - f(\tilde{A})) \le N_2(A - \tilde{A}) \sum_{j,k=0}^{\infty} \eta_{j,k} g_I^j(A) g_I^k(\tilde{A}),$$

provided the conditions (3.1) and $A - \tilde{A} \in SN_2$ hold.

If A and $\tilde{A}$ are normal operators, then Theorem 13.2 implies the inequality

$$N_2(f(A)B - Bf(\tilde{A})) \le N_2(K) \sup_{z \in co\,(A,\tilde{A})} |f'(z)|.$$

If $A \in SN_2$, then due to Theorem 3.1 one can replace $g_I(A)$ by $g(A)$, where

$$g(A) := \left[N_2^2(A) - \sum_{k=1}^{\infty} |\hat{\lambda}_k(A)|^2 \right]^{1/2}.$$

Similarly, if $\tilde{A} \in SN_2$, then one can replace and $g_I(\tilde{A})$ by $g(\tilde{A})$.

Example 13.1. Let $f(A) = e^{At}, t \ge 0$. Then

$$\sup_{z \in co\,(A,\tilde{A})} \left| \frac{d^{k+j+1}e^{zt}}{dz^{k+j+1}} \right| = e^{\alpha t} t^{k+j+1} \quad (j, k = 0, 1, 2, ...; \, t \ge 0),$$

where

$$\alpha := \max\{\sup \Re\sigma(A), \sup \Re\sigma(\tilde{A})\}.$$

Thus,

$$\eta_{j,k} = \frac{e^{\alpha t} t^{k+j+1}}{\sqrt{k!j!}(k+j+1)!} \quad (j, k = 0, 1, 2, ...).$$

Under conditions (3.1), (3.2), due to Theorem 13.2 we can write

$$N_2(e^{At}B - Be^{\tilde{A}t}) \le e^{\alpha t} N_2(K) \sum_{j,k=0}^{\infty} \frac{t^{k+j+1} g_I^k(A) g_I^j(\tilde{A})}{\sqrt{k!j!}(k+j+1)!} \quad (t \ge 0).$$

13.4 Perturbations of entire Banach valued functions

Let $\mathcal{X}$ and $\mathcal{Y}$ be complex normed spaces with norms $\|.\|_{\mathcal{X}}$ and $\|.\|_{\mathcal{Y}}$, respectively, and F a $\mathcal{Y}$-valued function defined on $\mathcal{X}$. Assume that $F(C + \lambda \tilde{C})$ $(\lambda \in \mathbf{C})$ is an entire function for all $C, \tilde{C} \in \mathcal{X}$. That is, for any $\phi \in \mathcal{Y}^*$, the functional $< \phi, F(C + \lambda \tilde{C}) >$ is an entire function.

Lemma 13.5. *Let $F(C + \lambda \tilde{C})$ $(\lambda \in \mathbf{C})$ be an entire function for all $C, \tilde{C} \in \mathcal{X}$ and there be a monotone non-decreasing function $G : [0, \infty) \to [0, \infty)$, such that*

$$\|F(C)\|_{\mathcal{Y}} \le G(\|C\|_{\mathcal{X}}). \tag{4.1}$$

Then

$$\|F(C) - F(\tilde{C})\|_{\mathcal{Y}} \le \|C - \tilde{C}\|_{\mathcal{X}} \, G(1 + \frac{1}{2}\|C + \tilde{C}\|_{\mathcal{X}} + \frac{1}{2}\|C - \tilde{C}\|_{\mathcal{X}}). \tag{4.2}$$

Proof. Put

$$Z_1(\lambda) = F(\frac{1}{2}(C + \tilde{C}) + \lambda(C - \tilde{C})).$$

Then $Z_1(\lambda)$ is an entire function and

$$F(C) - F(\tilde{C}) = Z_1(\frac{1}{2}) - Z_1(-\frac{1}{2}).$$

Thanks to the Cauchy integral,

$$Z_1(1/2) - Z_1(-1/2) = \frac{1}{2\pi i} \int_{|z|=1/2+r} Z_1(z) \frac{dz}{(z - 1/2)(z + 1/2)} \quad (r > 0).$$

Hence,

$$\|Z_1(1/2) - Z_1(-1/2)\|_{\mathcal{Y}} \le (1/2 + r) \sup_{|z|=1/2+r} \frac{\|Z_1(z)\|_{\mathcal{Y}}}{|z^2 - 1/4|} \le$$

$$\frac{1}{r} \sup_{|z|=1/2+r} |Z_1(z)|. \tag{4.3}$$

In addition, by (4.1)

$$\|Z_1(z)\|_{\mathcal{Y}} = \left\| F\left(\frac{1}{2}(C + \tilde{C}) + z(C - \tilde{C})\right) \right\|_{\mathcal{Y}}$$

$$\le G\left(\|\frac{1}{2}(C + \tilde{C}) + z(C - \tilde{C})\|_{\mathcal{X}}\right) \le$$

$$G\left(\frac{1}{2}\|C + \tilde{C}\|_{\mathcal{X}} + \left(\frac{1}{2} + r\right)\|C - \tilde{C}\|_{\mathcal{X}}\right) \quad (|z| = 1/2 + r).$$

Therefore according to (4.3),

$$\|F(C) - F(\tilde{C})\|_{\mathcal{Y}} = \|Z_1(1/2) - Z_1(-1/2)\|_{\mathcal{X}} \le$$
$$\frac{1}{r} G\left(\frac{1}{2}\|C + \tilde{C}\|_{\mathcal{X}} + (\frac{1}{2} + r)\|C - \tilde{C}\|_{\mathcal{X}}\right).$$

Taking

$$r = \frac{1}{\|C - \tilde{C}\|_{\mathcal{X}}},$$

we get the required result. Q. E. D.

13.5 Conservation of stability

Let $A, \tilde{A} \in \mathcal{B}(\mathcal{X})$. We will say that A is stable (Hurwitzian), if $\alpha(A) < 0$. Since A is bounded, it is simple to check that its stability is equivalent to the condition

$$u(A) := \int_0^\infty \|e^{At}\| dt < \infty.$$

We have

$$e^{At} - e^{\tilde{A}t} = \int_0^t e^{A(t-s)}[A - \tilde{A}]e^{\tilde{A}s} ds.$$

Hence,

$$\|e^{At} - e^{\tilde{A}t}\| \le \int_0^t \|e^{A(t-s)}\| \|A - \tilde{A}\| \|e^{\tilde{A}s}\| ds, \qquad (5.1)$$

and consequently,

$$\|e^{\tilde{A}t}\| \le \|e^{At}\| \int_0^t \|e^{A(t-s)}\| \|A - \tilde{A}\| \|e^{\tilde{A}s}\| ds. \qquad (5.2)$$

Let us investigate perturbations in the case when $\|\tilde{A}E - EA\|$ is small enough. Here $E = \tilde{A} - A$. Assume that A is stable and put

$$v_A = \int_0^\infty t\|e^{At}\| dt.$$

Theorem 13.3. *Let* A *be stable, and*

$$\|\tilde{A}E - EA\| v_A < 1. \qquad (5.3)$$

Then $\tilde{A}$ *is also stable. Moreover,*

$$u(\tilde{A}) \le \frac{u(A) + v_A \|E\|}{1 - v_A \|\tilde{A}E - EA\|} \qquad (5.4)$$

and

$$\int_0^\infty \|e^{\tilde{A}t} - e^{At}\| dt \le \|E\| v_A + \frac{\|\tilde{A}E - EA\| v_A(u(A) + v_A \|E\|)}{1 - v_A \|\tilde{A}E - EA\|}. \qquad (5.5)$$

To prove this result we need the following simple lemma.

Lemma 13.6. *Let $f(t)$, $u(t)$ and $v(t)$ be operator-valued functions defined on a finite segment $[a, b]$ of the real axis. Assume that $f(t)$ and $v(t)$ are boundedly differentiable and $u(t)$ is integrable on $[a, b]$. Then with the notation*

$$j_u(t) = \int_a^t u(s)ds \ (a < t \leq b),$$

the equality

$$\int_a^t f(s)u(s)v(s)ds = f(t)j_u(t)v(t)$$

$$- \int_a^t [f'(s)j_u(s)v(s) + f(s)j_u(s)v'(s)]ds$$

is valid.

Proof. Clearly,

$$\frac{d}{dt}f(t)j_u(t)v(t) = f'(t)j_u(t)v(t) + f(t)u(t)v(t) + f(t)j_u(t)v'(t).$$

Integrating this equality and taking into account that $j_u(a) = 0$, we arrive at the required result. Q. E. D.

By this lemma

$$e^{\tilde{A}t} - e^{At} = \int_0^t e^{\tilde{A}(t-s)} E e^{As}ds =$$

$$Ete^{At} + \int_0^t e^{\tilde{A}(t-s)}[\tilde{A}E - EA]se^{As}ds.$$

Hence,

$$\int_0^\infty \|e^{\tilde{A}t} - e^{At}\|dt \leq \int_0^\infty \|Ete^{At}\|dt$$

$$+ \int_0^\infty \int_0^t \|e^{\tilde{A}(t-s)}\|\|\tilde{A}E - EA\|\|se^{As}\|ds \, dt.$$

But

$$\int_0^\infty \int_0^t \|e^{\tilde{A}(t-s)}\|s\|e^{As}\|ds \, dt$$

$$= \int_0^\infty \int_s^\infty \|e^{\tilde{A}(t-s)}\| s \|e^{As}\| dt\, ds$$

$$= \int_0^\infty s\|e^{As}\| ds \int_0^\infty \|e^{\tilde{A}t}\| dt = v_A u(\tilde{A}).$$

Thus

$$\int_0^\infty \|e^{\tilde{A}t} - e^{At}\| dt \le \|E\| v_A + \|\tilde{A}E - EA\| v_A u(\tilde{A}). \tag{5.6}$$

Hence,

$$u(\tilde{A}) \le u(A) + \|E\| v_A + \|\tilde{A}E - EA\| v_A u(\tilde{A}).$$

So according to (5.3), we get (5.4). In addition, due to (5.6) and (5.4) we get (5.5). As claimed. Q. E. D.

Note also that the results presented in this section can be applied to the Dunford type operator functions. Recall that a Dunford type operator function $h(A)$ with a stable operator $-A$ is defined by

$$h(A) = \int_0^\infty e^{-At} f(t) dt,$$

where f is a scalar function and $h(z)$ is the Laplace transform to f. In particular,

$$A^{-1} = \int_0^\infty e^{-At} dt.$$

To illustrate the latter theorem, assume that $\mathcal{X} = \mathbf{C}^n$. Then making use of Example 3.2, we obtain $u(A) \le u_0(A)$ and $v_A \le \hat{v}_A$, where

$$u_0(A) := \sum_{k=0}^{n-1} \frac{g^k(A)}{|\alpha(A)|^{k+1}(k!)^{1/2}} \quad \text{and} \quad \hat{v}_A := \sum_{k=0}^{n-1} \frac{(k+1)g^k(A)}{|\alpha(A)|^{k+2}(k!)^{1/2}}.$$

Thus, Theorem 13.3 implies

Corollary 13.3. *Let $A, \tilde{A} \in \mathbf{C}^{n \times n}$, A be stable and $\|\tilde{A}E - EA\|\hat{v}_A < 1$. Then $\tilde{A}$ is also stable. Moreover,*

$$u(\tilde{A}) \le \frac{u_0(A) + \hat{v}_A\|E\|}{1 - \hat{v}_A\|\tilde{A}E - EA\|}$$

and

$$\int_0^\infty \|e^{\tilde{A}t} - e^{At}\| dt \le \|E\|\hat{v}_A + \frac{\|\tilde{A}E - EA\|\hat{v}_A(u_0(A) + \hat{v}_A\|E\|)}{1 - \hat{v}_A\|\tilde{A}E - EA\|}.$$

13.6   Comments to Chapter 13

Sections 13.1-13.4 are adopted from the paper [33]. The material in Section 13.5 is probably new.

Chapter 14

Functions of Two Non-commuting Operators in Hilbert Spaces

This chapter is devoted to a class of functions of two non-commuting operator arguments. For these functions, norm estimates are derived. Applications of the obtained estimates to the generalized polynomial equations in Hilbert spaces are also discussed.

14.1 Statement of the result

Let $\mathcal{H}$, $\mathcal{E}$ be separable Hilbert spaces, $A \in \mathcal{B}(\mathcal{H})$, $B \in \mathcal{B}(\mathcal{E})$ and $C \in \mathcal{B}(\mathcal{E}, \mathcal{H})$. By $co(A)$ we again denote the closed convex hull of $\sigma(A)$. Let Ω_A and Ω_B be neighborhoods of $co(A)$ and $co(B)$, respectively, $f(z, w)$ be a scalar function holomorphic on $\Omega_A \times \Omega_B$ and

$$F(f, A, C, B) := -\frac{1}{4\pi^2} \int_{\Gamma_B} \int_{\Gamma_A} f(z, w) R_z(A) C R_w(B) dw\, dz, \qquad (1.1)$$

where $\Gamma_A \subset \Omega_A, \Gamma_B \subset \Omega_B$ are closed Jordan contours surrounding $\sigma(A)$ and $\sigma(B)$, respectively.

It is assumed that

$$A_I = (A - A^*)/2i \in SN_2, B_I \in SN_2 \qquad (1.2)$$

and

$$C \in SN_2. \qquad (1.3)$$

Recall that

$$g_I(A) := \sqrt{2}[N_2^2(A_I) - \sum_{k=1}^{\infty} (\Im\, \lambda_k(A))^2]^{1/2}.$$

Let $\psi_{jk} = \psi_{jk}(f, A, B)$ be the numbers defined by

$$\psi_{00} = \sup_{z \in \sigma(A), w \in \sigma(B)} |f(z, w)|,$$

$$\psi_{jk} = \frac{1}{(j!k!)^{3/2}} \sup_{z\in co(A), w\in co(B)} \left| \frac{\partial^{j+k} f(z,w)}{\partial z^j \partial w^k} \right|,$$

$$\psi_{0k} := \frac{1}{(k!)^{3/2}} \sup_{z\in \sigma(A), w\in co(B)} \left| \frac{\partial^k f(z,w)}{\partial w^k} \right|$$

and

$$\psi_{j0} := \frac{1}{(j!)^{3/2}} \sup_{z\in co(A), w\in \sigma(B)} \left| \frac{\partial^j f(z,w)}{\partial z^j} \right| \quad (j,k = 1,2,...).$$

Theorem 14.1. *Let conditions (1.2) and (1.3) hold. If, in addition,* $f(z,w)$ *is regular on a neighborhood of* $co(A) \times co(B)$*, then*

$$\|F(f,A,C,B)\| \le N_2(C) \sum_{j,k=0}^{\infty} \psi_{jk} g_I^j(A) g_I^k(B).$$

If A *is normal,* B *non-normal, and* $f(z,w)$ *is regular on a neighborhood of* $\sigma(A) \times co(B)$*, then*

$$\|F(f,A,C,B)\| \le N_2(C) \sum_{k=0}^{\infty} \psi_{0k} g_I^k(B).$$

Similarly, if B *is normal,* A *non-normal, and* $f(z,w)$ *is regular on a neighborhood of* $co\,(A) \times \sigma(B)$*, then*

$$\|F(f,A,C,B)\| \le N_2(C) \sum_{k=0}^{\infty} \psi_{j0} g_I^j(A).$$

If both A *and* B *are normal and* $f(z,w)$ *is regular on a neighborhood of* $\sigma(A) \times \sigma(B)$*, then*

$$\|F(f,A,C,B)\| \le N_2(C) \sup_{z\in \sigma(A), w\in \sigma(B)} |f(z,w)|.$$

This theorem is proved in the next section.

14.2 Proof of Theorem 14.1

Due to Corollary 10.1, there is a sequence A_n ($n = 1,2,...$) of finite dimensional operators strongly converging to A, and a sequence of orthogonal finite dimensional projections $Z_n \xrightarrow{s} I$, such that $\sigma(A_n/Z_n\mathcal{H}) \subseteq \sigma(A)$, $A_n = A_n Z_n = Z_n A_n$ and $Z_n R_\lambda(A_n) = R_\lambda(A_n) Z_n \xrightarrow{s} R_\lambda(A)$ for all $\lambda \notin \sigma(A)$. Similarly, there is a sequence B_n of finite dimensional operators strongly converging to B, and a sequence of orthogonal finite dimensional projections $\tilde{Z}_n \overset{s}{\to} I$ such that $\sigma(B_n/\tilde{Z}_n\mathcal{H}) \subseteq \sigma(B)$, and $R_\lambda(B_n)\tilde{Z}_n = \tilde{Z}_n R_\lambda(B_n) \overset{s}{\to} R_\lambda(B)$ for and all $\lambda \notin \sigma(B)$. Consequently,

$$Z_n R_z(A_n) C R_w(B_n)\tilde{Z}_n \overset{s}{\to} R_z(A) C R_w(B) \quad (z \in \Gamma_A, w \in \Gamma_B).$$

By (1.1) we have

$$\Delta_n := F(f, A, C, B) - Z_n F(f, A_n, C, B_n)\tilde{Z}_n =$$

$$-\frac{1}{4\pi^2} \int_{\Gamma_B} \int_{\Gamma_A} f(z, w)(R_z(A) C R_w(B) - Z_n R_z(A_n) C R_w(B_n)\tilde{Z}_n)dw\, dz$$

Hence,

$$\|\Delta_n x\| \le \frac{1}{4\pi^2} \int_{\Gamma_B} \int_{\Gamma_A} |f(z, w)| h_n(z, w, x)dw\, dz \quad (x \in \mathcal{H}),$$

where

$$h_n(z, w, x) = \|(R_z(A) C R_w(B) - Z_n R_z(A_n) C R_w(B_n)\tilde{Z}_n)x\| \to 0.$$

In addition, $h_n(z, w, x)|f(z, w)|$ is uniformly bounded in n on $\Gamma_A \times \Gamma_B$. Thus by the Lebesgue theorem $\|\Delta_n x\| \to 0$. Hence, according to Theorem 1.1,

$$\|F(f, A, C, B)\| \le \sup_n \|F(f, A_n, C, B_n)\|. \tag{2.1}$$

Applying Theorem 4.3, we can write

$$\|Z_n F(f, A_n, C, B_n)\tilde{Z}_n\| \le N_2(C) \sum_{j=0}^{n-1} \sum_{k=0}^{\tilde{n}-1} \psi_{jk} g^j(A_n) g^k(B_n),$$

where $n = \operatorname{rank} A_n$, $\tilde{n} = \operatorname{rank} B_n$. Due to Theorem 3.1 and Corollary 10.2, $g(A_n) = g_I(A_n) \to g_I(A)$. Now inequality (2.1) yields the required result. Q. E. D.

14.3 Generalized polynomial equations in Hilbert spaces

Consider the equation

$$\sum_{j=0}^{m_1} \sum_{k=0}^{m_2} c_{jk} A^j X B^k = C \quad (m_1, m_2 < \infty), \tag{3.1}$$

where c_{jk} are complex numbers; $A \in \mathcal{B}(\mathcal{H})$, $B \in \mathcal{B}(\mathcal{E})$ and $C \in \mathcal{B}(\mathcal{E}, \mathcal{H})$, again. Denote

$$P(z, w) := \sum_{j=0}^{m_1} \sum_{k=0}^{m_2} c_{jk} z^j \tilde{w}^k \quad (z, w \in \mathbf{C})$$

and assume that

$$P(z, w) \neq 0 \quad ((z, w) \in co(A) \times co(B)). \tag{3.2}$$

Then by Theorem 2.1 equation (3.1) has a unique solution which is presentable by the formula

$$X = F\left(\frac{1}{P(z, w)}, A, C, B\right).$$

In this case $\psi_{jk} = \psi_{jk}(P)$ where

$$\psi_{00}(P) = \frac{1}{\inf_{z \in \sigma(A), w \in \sigma(B)} |P(z, w)|},$$

$$\psi_{jk}(P) = \frac{1}{(j!k!)^{3/2}} \sup_{z \in co(A), w \in co(B)} \left| \frac{\partial^{j+k}}{\partial z^j \partial w^k} \left(\frac{1}{P(z, w)}\right) \right|,$$

$$\psi_{0k}(P) = \frac{1}{(k!)^{3/2}} \sup_{z \in \sigma(A), w \in co(B)} \left| \frac{\partial^k}{\partial w^k} \left(\frac{1}{P(z, w)}\right) \right|,$$

and

$$\psi_{j0}(P) = \frac{1}{(j!)^{3/2}} \sup_{z \in co(A), w \in \sigma(B)} \left| \frac{\partial^j}{\partial z^j} \left(\frac{1}{P(z, w)}\right) \right| \quad (j, k \geq 1).$$

Now Theorem 14.1 implies

Corollary 14.1. *Let conditions (1.2), (1.3) and (3.2) hold. Then the unique solution X of (3.1) is subject to the inequality*

$$\|X\| \leq N_2(C) \sum_{j,k=0}^{\infty} \psi_{jk}(P) g_I^j(A) g_I^k(B).$$

If, in addition, A is normal, then

$$\|X\| \leq N_2(C) \sum_{j=0}^{\infty} \psi_{0j}(P) g_I^j(B).$$

If both A and B are normal, then

$$\|X\| \leq \frac{N_2(C)}{\inf_{z \in \sigma(A), w \in \sigma(B)} |P(z, w)|}.$$

Remark 14.1. Simple calculations show that if A is normal, then condition (3.2) can be replaced by $P(z, w) \neq 0$ $((z, w) \in \sigma(A) \times co(B))$. If both A and B are normal, then condition (3.2) can be replaced by

$$P(z, w) \neq 0 \quad ((z, w) \in \sigma(A) \times \sigma(B)).$$

14.4 Comments to Chapter 14

This chapter is based on the paper [47].

Bibliography

[1] Ahiezer, N.I. and Glazman, I.M. (1981). *Theory of Linear Operators in a Hilbert Space* (Pitman Advanced Publishing Program, Boston).

[2] Arendt, W., Batty, C., Neubrander, F. and Hieber, M. (2011). *Vector-valued Laplace Transforms and Cauchy Problems* (Birkhäuser, Basel).

[3] Baranov, N.I. and Brodskii, M.S. (1982). Triangular representations of operators with a unitary spectrum, *Funktsional. Anal. i Prilozhen.* **16**, 1, pp. 58–59. (Russian). English translation: *Functional Anal. Appl.* **16** (1982), 1, pp. 45–46.

[4] Baumgartel, H. (1985). *Analytic Perturbation Theory for Matrices and Operators*. Operator Theory, Advances and Appl., Vol. 52 (Birkhauser Verlag, Basel, Boston, Stuttgart).

[5] Bhatia, R. (1997). *Matrix Analysis* (Springer, New York).

[6] Bhatia, R. (2007). *Perturbation Bounds for Matrix Eigenvalues*, Classics in Applied Mathematics, Vol. 53 (SIAM, Philadelphia, PA).

[7] Bhatia, R. and Rosenthal, P. (1997). How and why to solve the matrix equation $AX - XB = Y$, *Bull. London Math. Soc.*, **29**, pp. 1–21.

[8] Bhatia, R. and Uchiyama, M. (2009). The operator equation $A^{n-i}XB^i = Y$. *Expo. Math.* **27**, pp. 251–255.

[9] Branges, L. de. (1965). Some Hilbert spaces of analytic functions II, *J. Math. Analysis and Appl.*, **11**, pp. 44–72.

[10] Branges, L. de. (1965). Some Hilbert spaces of analytic functions III, *J. Math. Analysis and Appl.*, **12**, pp. 149–186.

[11] Brodskii, M.S. (1960). Triangular representation of some operators with completely continuous imaginary parts. *Dokl. Akad. Nauk SSSR*, **133**, pp. 1271–1274 (Russian). English translation: *Soviet Math. Dokl.* 1 (1960), pp. 952–955.

[12] Brodskii, M.S. (1971). *Triangular and Jordan Representations of Linear Operators*, Transl. Math. Mongr., Vol. 32, (Amer. Math. Soc., Providence, R. I.).

[13] Brodskii, V.M., Gohberg, I.C. and Krein M.G. (1969). General theorems on triangular representations of linear operators and multiplicative representations of their characteristic functions, *Funk. Anal. i Pril.*, **3**, pp. 1–27

(Russian). English translation: *Func. Anal. Appl.*, **3** (1969), pp. 255–276.

[14] Daleckii, Yu. L. and Krein, M.G. (1974). *Stability of Solutions of Differential Equations in Banach Space* (Amer. Math. Soc., Providence, R. I.).

[15] Davis, C. and Kahan, W. (1969). Some new bounds on perturbation of subspaces, *Bull. Amer. Math. Soc.* **75**, pp. 863–868.

[16] Davis, C. and Kahan, W. (1970). The rotation of eigenvectors by a perturbation III, *SIAM J. Numer. Anal.* **7**, pp. 1–46.

[17] Dunford, N. and Schwartz, J.T. (1963). *Linear Operators, part II. Spectral Theory* (Interscience, New York, London).

[18] Dunford, N. and Schwartz, J.T. (1966). *Linear Operators, part I* (Interscience, New York, London).

[19] Dunford, N. and Schwartz, J.T. (1971). *Linear Operators, part III, Spectral Operators* (Wiley-Interscience Publishers, Inc., New York).

[20] Gantmakher, F.R. (1967). *Theory of Matrices* (Nauka, Moscow). (In Russian).

[21] Gel'fand, I.M. and Shilov, G.E. (1958). *Some Questions of Theory of Differential Equations* (Nauka, Moscow). (In Russian).

[22] Gelfond, A.O. (1967). *Calculations of Finite Differences* (Nauka, Moscow). (In Russian).

[23] Gil', M.I. (1973). On the representation of the resolvent of a nonselfadjoint operator by the integral with respect to a spectral function, *Soviet Math. Dokl.*, **14**, pp. 1214–1217.

[24] Gil', M.I. (1979). An estimate for norms of resolvent of completely continuous operators, *Mathematical Notes*, **26**, pp. 849–851.

[25] Gil', M.I. (1979). Estimating norms of functions of a Hilbert-Schmidt operator (in Russian), *Izvestiya VUZ, Matematika*, **23**, pp. 14–19. English translation in *Soviet Math.*, **23** (1979), pp. 113–119.

[26] Gil', M.I. (1983). One estimate for resolvents of nonselfadjoint operators which are "near" to selfadjoint and to unitary ones, *Mathematical Notes*, **33**, pp. 81–84.

[27] Gil', M.I. (1998). *Stability of Finite and Infinite Dimensional Systems* (Kluwer Academic Publishers, Boston).

[28] Gil', M.I. (2003). *Operator Functions and Localization of Spectra*, Lecture Notes in Mathematics, Vol. 1830, (Springer-Verlag, Berlin).

[29] Gil', M.I. (2005). *Explicit Stability Conditions for Continuous Systems*, Lecture Notes In Control and Information Sciences, Vol. 314 (Springer Verlag, Berlin).

[30] Gil', M.I. (2007). *Difference Equations in Normed Spaces. Stability and Oscillations*, North-Holland, Mathematics Studies, Vol. 206 (Elsevier, Amsterdam).

[31] Gil', M.I. (2008). Spectrum and resolvent of a partial integral operator. *Appl. Anal.* **87**, no. 5, pp. 555–566.

[32] Gil', M.I. (2008). Inequalities of the Carleman type for Neumann-Schatten operators, *Asian-European J. of Math.*, **1**, pp. 203–212.

[33] Gil', M.I. (2010). Perturbations of functions of operators in a Banach space. *Math. Phys. Anal. Geom.* **13**, no. 1, pp. 69–82.

[34] Gil', M.I. (2010). *Localization and Perturbation of Zeros of Entire Functions.* Lecture Notes in Pure and Applied Mathematics, Vol. 258 (CRC Press, Boca Raton, FL).

[35] Gil', M.I. (2013). *Stability of Vector Differential Delay Equations* (Birkhäuser Verlag, Basel).

[36] Gil, M.I. (2014). *Stability of Neutral Functional Differential Equations*, Atlantis Studies in Differential Equations, Vol. 3 (Paris, France).

[37] Gil, M.I. (2014). Norm estimates for solutions of matrix equations $AX - XB = C$ and $X - AXB = C$, *Discussiones Mathematicae. Differential Inclusions Control and Optimization* **34(2)**, pp. 35-47.

[38] Gil, M.I. (2014). A bound for condition numbers of matrices, *Electronic Journal of Linear Algebra*, **27**, pp. 162-171.

[39] Gil, M.I. (2014). Resolvents of operators inverse to Schattenvon Neumann ones, *Ann. Univ. Ferrara Sez. VII Sci. Mat.* **60**, 2, pp. 363–376.

[40] Gil', M.I. (2014). A bound for similarity condition numbers of unbounded operators with Hilbert-Schmidt hermitian components, *J. Aust. Math. Soc.* **97**, pp. 331–342.

[41] Gil', M.I. (2015). On stability of linear Barbashin type integro-differential equations, *Mathematical Problems in Engineering*, **2015**, Article ID 962565, 5 pages.

[42] Gil', M.I. (2015). On condition numbers of spectral operators in a Hilbert space, *Anal. Math. Phys.*, **5**, pp. 363–372.

[43] Gil', M.I. (2015). Perturbations of invariant subspaces of operators with Hilbert-Schmidt Hermitian Components, *Arch. Math. (Basel)* **105**, 5, pp. 447–452.

[44] Gil', M.I. (2015). Norm estimates for solutions of polynomial operator equations, *Journal of Mathematics*, **2015**, Article ID 524829, 7 pages.

[45] Gil', M.I. (2015). Exponential stability of nonlinear non-autonomous multivariable systems, *Discussiones Mathematicae. Differential Inclusions, Control and Optimization*, **35(1)**, pp. 1–12.

[46] Gil', M.I. (2015). Stability of linear nonautonomous multivariable systems with differentiable matrices, *Systems & Control Letters*, **81**, pp. 31–33.

[47] Gil, M.I. (2015). Norm estimates for functions of two non-commuting operators, *Rocky Mountain Journal of Mathematics* **45**, no. 2, pp. 927–939.

[48] Gil', M.I. (2016). Resolvents of operators on tensor products of Euclidean spaces, *Linear and Multilinear Algebra*, **64(4)**, pp. 699–716.

[49] Gil', M.I. (2016). Perturbations of invariant subspaces of compact operators *Acta Scientiarum Mathematicarum (Szeged)*, **82**, 1-2, pp. 271–279.

[50] Gil', M.I. (2016). Stability and boundedness of solutions to nonautonomous parabolic integrodifferential equations, *Mathematical Problems in Engineering*, **2016**, Article ID 2468458, 6 pages.

[51] Gohberg, I.C., Goldberg, S. and Kaashoek M.A. (1990). *Classes of Linear Operators, Vol. 1*, (Birkhäuser Verlag, Basel).

[52] Gohberg, I.C., Goldberg, S. and Kaashoek M.A. (1993). *Classes of Linear Operators, Vol. 2*, (Birkhäuser Verlag, Basel).

[53] Gohberg, I.C., Goldberg, S. and Krupnik, N. (2000). *Traces and Determi-*

nants of Linear Operators (Birkhäuser Verlag, Basel).

[54] Gohberg, I.C. and Krein, M.G. (1969). *Introduction to the Theory of Linear Nonselfadjoint Operators*, Trans. Mathem. Monographs, Vol. 18 (Amer. Math. Soc., R. I).

[55] Gohberg, I.C. and Krein, M.G. (1970). *Theory and Applications of Volterra Operators in a Hilbert Space*, Trans. Mathem. Monographs, Vol. 24 (Amer. Math. Soc., R. I).

[56] Graham, A. (1981). *Kronecker Product and Matrix Calculus with Applications* (Ellis Horwood Limited, Chichester).

[57] Grubisic, L., Truhar, N. and Veseli, K. (2012). The rotation of eigenspaces of perturbed matrix pairs, *Linear Algebra and its Applications* **436**, pp. 4161–4178.

[58] Halmos, P. (1982). *Hilbert Space Problem Book*, Second edition (Springer Verlag, New York).

[59] Horn, R.A. and Johnson Ch. R. (1991). *Topics in Matrix Analysis* (Cambridge University Press, Cambridge).

[60] Kato, T. (1980). *Perturbation Theory for Linear Operators* (Springer-Verlag, Berlin).

[61] Konstantinov, M., Gu, D.-W., Mehrmann, V. and Petkov, P. (2003). *Perturbation Theory for Matrix Equations*. Studies in Computational Mathematics, Vol. 9 (North Holland, Amsterdam).

[62] Krein, S.G. (1971). *Linear Equations in a Banach Space* (Amer. Math. Soc., Providence, R.I).

[63] Krein, S.G. (1982). *Functional Analysis* (Birkhauser, Boston).

[64] Levin, B. Ya. (1996). *Lectures on Entire Functions*, Trans. of Math. Monographs, Vol. 150 (Amer. Math. Soc., R.I).

[65] Locker, J. (1999). *Spectral Theory of Nonselfadjoint Two Point Differential Operators*. Mathematical Surveys and Monographs, Volume 73 (Amer. Math. Soc, R.I).

[66] Lyubich, Yu. I. and Macaev, V.I. (1962). On operators with separable spectrum, *Mathem. Sbornik*, **56 (98)**, 4, pp. 433–468 (Russian). English translation: *Amer. Math. Soc. Transl* (Series 2), **47** (1965), pp. 89–129.

[67] Macaev, V.I. (1961). A class of completely continuous operators, *Dokl. Akad. Nauk SSSR* **139 (2)**, pp. 548–551 (Russian); English translation: *Soviet Math. Dokl.* **1**, (1961), pp. 972–975.

[68] Marcus, M. and Minc, H. (1964). *A Survey of Matrix Theory and Matrix Inequalities* (Allyn and Bacon, Boston).

[69] Martínez, C. and Sanz, M. (2001). *The Theory of Fractional Powers of Operators*. North-Holland Mathematics Studies, 187 (North-Holland Publishing Co., Amsterdam).

[70] Mazko, A.G. (2008). *Matrix Equations, Spectral Problems and Stability of Dynamic Systems. Stability, Oscillations and Optimization of Systems* (Scientific Publishers, Cambridge).

[71] Ostrowski, A.M. (1973). *Solution of Equations in Euclidean and Banach Spaces* (Academic Press, New York - London).

[72] Pazy, A. (1983). *Semigroups of Linear Operators and Applications to Partial*

Differential Equations (Springer-Verlag, New York).

[73] Pietsch, A. (1987). *Eigenvalues and s-numbers* (Cambridge University Press, Cambridge).

[74] Radjavi, H. and Rosenthal, P. (1973). *Invariant Subspaces* (Springer-Verlag, Berlin).

[75] Rosenblum, M. (1956). On the operator equation $BX - XA = Q$, *Duke Math. J.* **23**, pp. 263–270.

[76] Stewart, G.W. and Sun, J.G. (1990). *Matrix Perturbation Theory* (Academic Press, New York).

List of Symbols

$(.,.)$ scalar product

$[A, B] = AB - BA$ commutator of A and B

$\|A\|$ operator norm of an operator A

$|A| = |A|_e$ matrix whose elements are absolute values
    of matrix A in its Schur basis

A^{-1} inverse to A

A^* conjugate to A

$A_I = \Im A = (A - A^*)/2i$

$A_R = \Re A = (A + A^*)/2$

$\alpha(A) = \sup Re\ \sigma(A)$

$\mathcal{B}(\mathcal{X})$ algebra of bounded linear operators in $\mathcal{X}$

$\mathcal{B}(\mathcal{Y}, \mathcal{X})$ set of bounded operators acting from $\mathcal{Y}$ to $\mathcal{X}$

$\mathbf{C}^n$ complex Euclidean space

$\det(A)$ determinant of A

$g(A)$ is defined on page 43

$g_I(A)$ is defined on page 148

$\mathcal{H}$ Hilbert space

$I = I_{\mathcal{X}}$ identity operator (in space $\mathcal{X}$)

$\lambda_k(A)$ eigenvalue of A

$N_p(A)$ Schatten-von Neumann norm of A

$N_2(A)$ Hilbert-Schmidt (Frobenius) norm of A

$\mathcal{P}$ a maximal chain of projections

$\mathbf{R}^n$ real Euclidean space

$R_\lambda(A)$　　　resolvent of A

$r_s(A)$　　　spectral radius of A

$r_{\text{low}}(A)$　　　lower spectral radius of A

$\rho(A, \lambda)$　　　distance between $\lambda \in \mathbf{C}$ and the spectrum of A

SN_1　　　Trace class

SN_2　　　Hilbert-Schmidt ideal

SN_p　　　Schatten-von Neumann ideal

$s_j(A)$　　　s-number (singular number) of A

$sv_A(B)$　　　spectral variation of B with respect to A

$\xrightarrow{s}$　　　strong convergence

$\sigma(A)$　　　spectrum of A

$\mathcal{X}$　　　Banach space

Index

Printed in the United States
By Bookmasters